C0-FYM-291

WITHDRAWN

WILLIAM F. MAAG LIBRARY
YOUNGSTOWN STATE UNIVERSITY

Number 305

O. C. Garcia
and W. Taylor

The lattice of interpretability types of varieties

Memoirs
of the American Mathematical Society

Providence · Rhode Island · USA

July 1984 · Volume 50 · Number 305 (second of 3 numbers) · ISSN 0065-9266

color code: 182
lettering: black

Memoirs of the American Mathematical Society
Number 305

Y 10 2

O. C. Garcia
and W. Taylor

The lattice of interpretability types of varieties

Published by the
AMERICAN MATHEMATICAL SOCIETY
Providence, Rhode Island, USA

July 1984 · Volume 50 · Number 305 (second of 3 numbers)

MEMOIRS of the American Mathematical Society

This journal is designed particularly for long research papers (and groups of cognate papers) in pure and applied mathematics. It includes, in general, longer papers than those in the TRANSACTIONS.

Mathematical papers intended for publication in the Memoirs should be addressed to one of the editors. Subjects, and the editors associated with them, follow:

Ordinary differential equations, partial differential equations and applied mathematics to JOEL A. SMOLLER, Department of Mathematics, University of Michigan, Ann Arbor, MI 48109.

Complex and harmonic analysis to LINDA PREISS ROTHSCHILD, Department of Mathematics, University of California at San Diego, LaJolla, CA 92093

Abstract analysis to WILLIAM B. JOHNSON, Department of Mathematics, Ohio State University, Columbus, OH 43210

Algebra, algebraic geometry and number theory to LANCE W. SMALL, Department of Mathematics, University of California at San Diego, LaJolla, CA 92093

Logic, set theory and general topology to KENNETH KUNEN, Department of Mathematics, University of Wisconsin, Madison, WI 53706

Topology to WALTER D. NEUMANN, Department of Mathematics, University of Maryland, College Park, MD 20742

Global analysis and differential geometry to TILLA KLOTZ MILNOR, Department of Mathematics, University of Maryland, College Park, MD 20742

Probability and statistics to DONALD L. BURKHOLDER, Department of Mathematics, University of Illinois, Urbana, IL 61801

Combinatorics and number theory to RONALD GRAHAM, Mathematical Sciences Research Center, AT&T Bell Laboratories, 600 Mountain Avenue, Murray Hill, NJ 07974

All other communications to the editors should be addressed to the Managing Editor, R. O. WELLS, JR., Department of Mathematics, University of Colorado, Boulder, CO 80309

MEMOIRS are printed by photo-offset from camera-ready copy fully prepared by the authors. Prospective authors are encouraged to request booklet giving detailed instructions regarding reproduction copy. Write to Editorial Office, American Mathematical Society, P. O. Box 6248, Providence, Rhode Island 02940. For general instructions, see last page of Memoir.

SUBSCRIPTION INFORMATION. The 1984 subscription begins with Number 289 and consists of six mailings, each containing one or more numbers. Subscription prices for 1984 are $148 list; $74 member. A late charge of 10% of the subscription price will be imposed upon orders received from nonmembers after January 1 of the subscription year. Subscribers outside the United States and India must pay a postage surcharge of $10; subscribers in India must pay a postage surcharge of $15. Each number may be ordered separately; *please specify number* when ordering an individual number. For prices and titles of recently released numbers, refer to the New Publications sections of the NOTICES of the American Mathematical Society.

BACK NUMBER INFORMATION. For back issues see the AMS Catalogue of Publications.

TRANSACTIONS of the American Mathematical Society

This journal consists of shorter tracts which are of the same general character as the papers published in the MEMOIRS. The editorial committee is identical with that for the MEMOIRS so that papers intended for publication in this series should be addressed to one of the editors listed above.

Subscriptions and orders for publications of the American Mathematical Society should be addressed to American Mathematical Society, P. O. Box 1571, Annex Station, Providence, R. I. 02901. *All orders must be accompanied by payment.* Other correspondence should be addressed to P. O. Box 6248, Providence, R. I. 02940.

MEMOIRS of the American Mathematical Society (ISSN 0065-9266) is published bimonthly (each volume consisting usually of more than one number) by the American Mathematical Society at 201 Charles Street, Providence, Rhode Island 02904. Second Class postage paid at Providence, Rhode Island 02940. Postmaster: Send address changes to Memoirs of the American Mathematical Society, American Mathematical Society, P. O. Box 6248, Providence, RI 02940.

Copyright © 1984, American Mathematical Society

Information on Copying and Reprinting can be found at the back of this journal.

The paper used in this journal is acid-free and falls within the guidelines established to ensure permanence and durability.

TABLE OF CONTENTS

ABSTRACT

We investigate the lattice, invented by W. D. Neumann in 1974, formed by the class of all varieties under the quasi-ordering "V is interpretable in W." The lattice is found to be non-modular and a proper class. Various familiar varieties are found to be Λ-irreducible (or prime) and various filters (especially Mal'tsev classes) are found to be indecomposable (or prime).

Many familiar varieties are found to be inequivalent in the lattice, using a new technique of SIN algebras. Seven figures are included which document the known relationships between some sixty known or easily describable varieties and varietal families.

1980 Mathematics Subject Classification. Primary 08B05; Secondary 18C10, 22A99, 06B99.

Library of Congress Cataloging in Publication Data

García, O. C. (Octavio Carlos), 1933-

The lattice of interpretability types of varieties.

(Memoirs of the American Mathematical Society, ISSN 0065-9266 ; no. 305 (July 1984))
Bibliography: p.
1. Algebra, Universal. 2. Lattice theory.
3. Equations, Theory of. 4. Category theory.
I. Taylor, W. (Walter), 1940- . II. Title.
III. Series: Memoirs of the American Mathematical Society ; no. 305. IV. Title: Interpretability types of varieties.
QA3.A57 no. 305 [QA251] 510 s [512] 84-10997

ISBN 0-8218-2308-6

Copying and reprinting. Individual readers of this publication, and nonprofit libraries acting for them, are permitted to make fair use of the material, such as to copy an article for use in teaching or research. Permission is granted to quote brief passages from this publication in reviews provided the customary acknowledgement of the source is given.

Republication, systematic copying, or multiple reproduction of any material in this publication (including abstracts) is permitted only under license from the American Mathematical Society. Requests for such permission should be addressed to the Executive Director, American Mathematical Society, P.O. Box 6248, Providence, Rhode Island 02940.

The owner consents to copying beyond that permitted by Sections 107 or 108 of the U.S. Copyright Law, provided that a fee of $1.00 plus $.25 per page for each copy be paid directly to Copyright Clearance Center, Inc., 21 Congress Street, Salem, Massachusetts 01970. When paying this fee please use the code 0065-9266/84 to refer to this publication. This consent does not extend to other kinds of copying, such as copying for general distribution, for advertising or promotion purposes, for creating new collective works or resale.

QA
3
.A57
no.305

ACKNOWLEDGEMENTS

The authors thank the National Science Foundation (USA) which partially supported Taylor's research under grants MCS 79-00907 and MCS 83-01065, and the Instituto de Matemáticas of UNAM (México) which supported García's stay in Colorado during parts of 1982 and 1983. Moreover, we thank the following mathematicians for various useful conversations and help in preparing this research: Alan Day, Ralph Freese, Don Pigozzi, Ralph McKenzie, and J. A. de la Peña. Very special thanks go to Jan Mycielski, whose continuing interest and encouragement has been a real inspiration. The first author, O. C. García, also expresses his gratitude to Walter Taylor for many comments concerning his contributions to this research. Both authors wish to thank Cathy Kosta for her expert typing of the final copy.

WILLIAM F. MAAG LIBRARY
YOUNGSTOWN STATE UNIVERSITY

INTRODUCTION

In 1974, W.D. Neumann introduced [63] an interesting lattice for comparing the relative strengths of varieties (or, one might say, equational theories). He defined a variety V to be $\leq$ another variety W (not necessarily of the same similarity type) iff V is *interpretable* in W. This means that for each V-operation $F_t(x_1, \ldots, x_n)$ there exists a term $\alpha_t(x_1, \ldots, x_n)$ in the language of W (i.e., a formal composition of some W-operations) such that if $(A; G_s)_{s \in S}$ is any algebra in W, then $(A; \overline{\alpha}_t) \in V$, where $\overline{\alpha}_t$ is the natural interpretation of the term α_t in the algebra $(A; G_s)$. The ordering $\leq$ defines only a quasi-ordering on the class of all varieties, but, as usual, a partial ordering is obtained if we identify V and W whenever $V \leq W \leq V$. The resulting quotient partial order turned out to be a lattice, which Neumann found to be useful for his study of Mal'tsev consitions (which are filters in this lattice).

This memoir is an extensive study of Neumann's lattice (which we will denote L), both from the theoretical point of view (what are its $\wedge$-prime elements?, etc.), and from the detailed "pictorial" point of view (can one get a good picture of how the more familiar varieties are situated in L?). Before outlining any results in this introduction, we feel it is important to state our convention regarding zero-ary operations: for us, the clone begins with unary operations, and so a "constant" can only be included in the form of $f(x)$ obeying $f(x) \approx f(y)$. A different lattice L' will result if one adopts the alternative convention that constants have an independent existence as the zero-layer of the clone, and that such a constant cannot legally be interpreted by a unary term $\alpha(x)$. Our choice of convention stems not from any special belief that it is "right," but simply from the facts that L is simpler than L' and that the structure of L' may be easily inferred from that of L, as follows. The lattice L has a least variety with a constant operation, namely the variety C defined by $fx \approx fy$. As the reader may easily check, every $V \geq C$ separates in L' into exactly two L'-inequivalent varieties, one with an empty algebra, and the other without. Thus, L' is formed from L by Alan Day's [19] construction of splitting the interval $[C,1]$. One can easily see that this splitting process introduces various covering relations, $\vee$-irreducibles, etc., into L', which were not already in L. In concentrating on

Received by the editor October 14, 1983.

L, we do not have to pay any attention to these spurious relationships; moreover, there is an obvious advantage in drawing a picture of L.

We wish to point out the affiliation of this lattice-theoretic perspective on variety theory with a similar lattice-theoretic perspective on logic and set theory, which was developed by Jan Mycielski around 1977 [60]. He introduced a definition of interpretability for arbitrary first order theories, which we will not repeat, except to point out two important regards in which it differs from ours. Firstly, his definition had $T \leq S$ as soon as every finite subtheory of T is $\leq S$. In our lattice, this so-called *local* interpretability would mean that $V \leq W$ as soon as $V_0 \leq W$ for every V_0 defined by finitely many of the axioms of V, which certainly fails in L as we have defined it. This alternate lattice, call it L", is related to our L in the following way: it is the lattice of ideals in L_0, the sublattice of L consisting of finitely based varieties. All the information in this memoir concerning finitely based varieties will also hold in L", but some of the more infinitary aspects (notably the proper classes we have found) will no longer hold for L".

The second important difference is that Mycielski's lattice is distributive, but ours is not even modular. The distributivity of the lattice associated with pure logic simply reflects the well known distributivity between "and" and "or." In our case, however, even though joins are formed by a sort of conjunction, there is no equational substitute for disjunction, and so meets take a subtler form in L, whereby distributivity is lost.

Having clearly specified the lattice L, we may continue the introduction by outlining how we will be describing L in the various sections of the memoir. In the first section we give a detailed presentation of finite joins and meets in L. (Although in a sense this is a review of material from [63], our presentation seems new to us.) This introductory section also contains a few basic facts about the lattice (e.g., a characterization of its 0 and 1), and a brief discussion of "k^{th} power varieties."

In §2 we provide a category-theoretic point of view and introduce some major landmarks of our lattice, namely the varieties V_κ = the variety generated by a κ-element set together will all finitary operations on that set. It is here that we establish that L is a *proper class* (since all V_κ are distinct.)

In §3 we use the spectrum to give some examples of failure of modularity in the lattice L.

In §4 we establish that various known varieties are $\wedge$-*prime* in L: Boolean algebra, lattice theory, semilattices, etc. (Group theory is known to be $\wedge$-irreducible, but $\wedge$-primeness is open.) More generally, we use the size of free algebras to estimate Helly numbers of a variety V in L. (We say V has Helly number $\geq \kappa$ if $V \geq V_0 \wedge \ldots \wedge V_\kappa$ implies that $V \geq$ some proper submeet.)

In §5 we consider some known and some lesser known filters in our lattice, proving along the way that some of them are prime. This includes a review of the theory of Mal'tsev conditions; along the way we introduce some heretofore unrecognized Mal'tsev conditions, such as:

$$\{V : V > \text{Group Theory}\}$$

Among the Mal'tsev filters which we prove to be prime are:

congruence-regularity
congruence-weak-uniformity
coherence.

Two examples of $\vee$-prime elements are 1 and $x + y \approx y + x$. It is unknown whether Mal'tsev's identity $pxyy \approx x$, $pxxy \approx y$ is $\vee$-prime. Likewise, it is unknown whether congruence-modularity defines a prime filter.

In §6 we prove that congruence-distributivity is not a prime filter (in fact, it is the intersection of two larger Mal'tsev filters). Moreover, we give a rather general condition for V to be $\vee$-reducible in L $(C < V < 1$, where C is the variety with one constant operation); a similar condition is given for filters to be not prime or not indecomposable. Thus, for example, group theory is a proper join in our lattice. (But there remains the intriguing possibility that each of the joinands is irreducible.)

In §7 we discuss k^{th} roots, and in §8, tensor products. Neither of these topics is strictly lattice-theoretic in its definition, but the definitions are *lattice-invariant*, thus yielding new structure on L. Here we prove that an important filter of J. R. Isbell is prime.

In §9 we begin a detailed description of L. In fact, the work of the last three sections (§§9-11) is best summarized by the seven figures which immediately follow this introduction. These figures depict the partial ordering of L, to the extent we understand it, especially for those varieties which are either already known or easily described. But an indication of the true lattice structure has not really been possible: most familiar varieties are $\wedge$-irreducible, and also seemingly not $\vee$-reducible except in unfamiliar ways.

Generally speaking, the hard part in §§9-11, i.e., the difficulty in preparing the diagrams, has been to establish the negative relations $V \not\leq W$. By contrast, the known instances of $V \leq W$ are almost all obvious, in that the terms α_t required for interpretation are readily apparent. Of course if V satisfies a certain Mal'tsev condition and W does not, then we automatically have $V \not\leq W$; but the known Mal'tsev conditions were not sufficient to separate many varieties; e.g., various varieties of groups. And so a large part of §§9-11 is devoted to new techniques for separation of varieties in L. The primary new technique is the consideration of SIN algebras ("simple under interpretation"), which are in a sense unchanged under interpretation; hence if such an algebra lies in W but not in V, then again we have $V \nleq W$. The precise definition and the theoretical development are in §1, but the detailed applications are in §§9-11. Of course the pioneering method for separating points of L lay in the detailed syntactical ("word problem") studies of K. Fichtner and D. Kelly from the early seventies (see [72] for references). We have occasionally used this technique (e.g., in Proposition 85), but never with a particularly difficult word problem. We caution the reader that this work is not complete; in some cases we know, say, $V \leq W$, but are unsure of $W \nleq V$. In such a case, our diagram has V below W with a *light line* joining them. A *dark line* indicates that we have proved $V < W$.

In one case, we found $V \leq W$ in a surprising and non-obvious way:

$$Ab_n =_L N_2 \cap B_n$$

for any odd n, where N_2 is the variety of 2-nilpotent groups, B_n is the (Burnside) variety of groups of exponent n, and Ab_n is the Abelian part of B_n. The relation $\geq$ is obvious, but for $\leq$ we need to define an Abelian group operation by the term

$$\alpha(x,y) := xy[x,y]^p$$

where $n = 2p + 1$. For the details, see Proposition 57 and the preceding Lemma.

In §9, we first obtain enough information about points in the lattice as a whole to be able to prepare the first Figure (which also summarizes some salient points from §§1-8). The remainder of §9 is concerned with varieties of groups. The associated Figures (2, 3 and 4) may thus be thought of as "blow-ups" of the L-interval [Gru,Ab].

Then §10 is concerned with the part of L below Ab = Abelian groups. Figure 5 gives a detailed picture of the interval [0,Ab] (especially the part below monoids), and in Figures 6 and 7 we view two subintervals of [0,Ab] with a scope of even higher resolution. One interesting discovery which cannot really be included in the figures is the prevalence of *proper classes* in L. For instance, since we originally thought of C (the variety with one constant operation) as rather "low" in L, we were thus surprised to discover that [0,C] is a proper class. (The points Q_a in Figure 5 represent an ordinal sequence below C; for a full explanation, see Proposition 83 on page 109.) (Another proper class is found in §11.) Moreover, both of the "small" intervals in Figures 6 and 7 are seen to have infinite breadth and height.

Among the easily described but surprising "points" in L are every ring (occurring as a variety of modules) and every topological space. To see how a space A occurs, we note that the class of varieties which have a topological model based on the space A form an ideal in L. As we will see in §2, this ideal is principal, and its generator is called V(A). Thus the location of V(A) in L is an indicator of how many different identities can be modeled by continuous operations on the space A, and as such, promises to form an interesting algebraic invariant of the space A. The second author is preparing a paper on this and related topics. In Figure 1, we have indicated the location of V(A) for A a closed interval (denoted •——•), and for A a circle (denoted O).

Among the many intriguing questions about L which remain unsettled at this writing, three stand out: does zero have a cover in L?, are there any covers in L?, and do Mal'tsev's identities $pxzz \approx x$, $pxxz \approx z$ form a $\bigvee$-irreducible or $\bigvee$-prime element of L? If L were known to be algebraic, then we could obtain a cover, but this remains unknown. We conclude the memoir with a list of open problems.

THE FIGURES

The following conventions are in force for the figures appearing on the next seven pages. A dictionary of all variety names appears right after the last figure.

All Figures: A *dark* segment rising from a to b means $a < b$. A *light* segment means only $a \leq b$ (with $b \leq a$ remaining open).

Figure 1: $\overline{0}$ - $\bigvee$-irreducible $\underline{0}$ - $\bigwedge$-irreducible

$\overline{\overline{0}}$ - $\bigvee$-prime $\underline{\underline{0}}$ - $\bigwedge$-prime

⦶ - not $\bigvee$-irreducible

⦷ - not $\bigvee$-prime

Figure 3: A medium thick segment indicates that our proof of $a < b$ is based on Proposition 51.

Figures 2,3,4: Juxtaposition of the names of two group varieties abbreviates their intersection.

The filters depicted in Figure 1 are those of congruence-distributivity and congruence-permutability. The ideal there is that of essentially unary varieties.

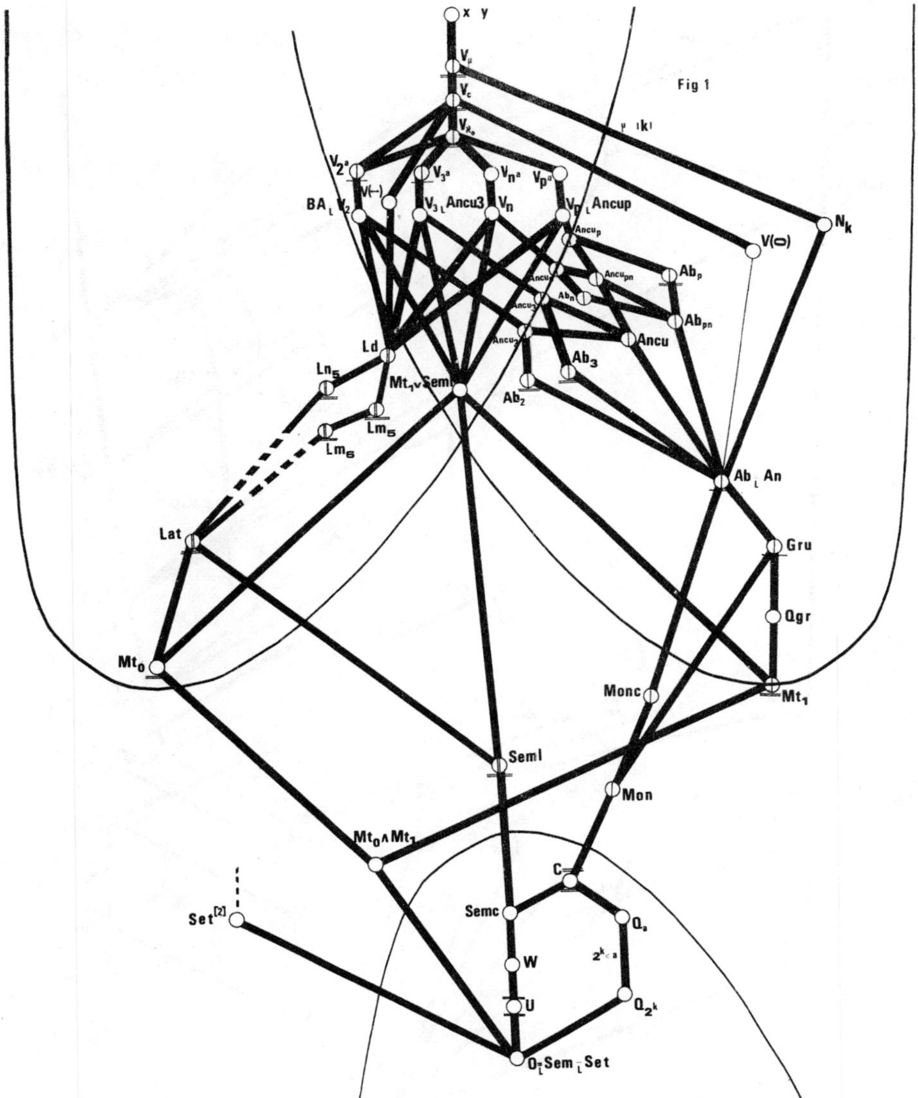

Fig 1

WILLIAM F. MAAG LIBRARY
YOUNGSTOWN STATE UNIVERSITY

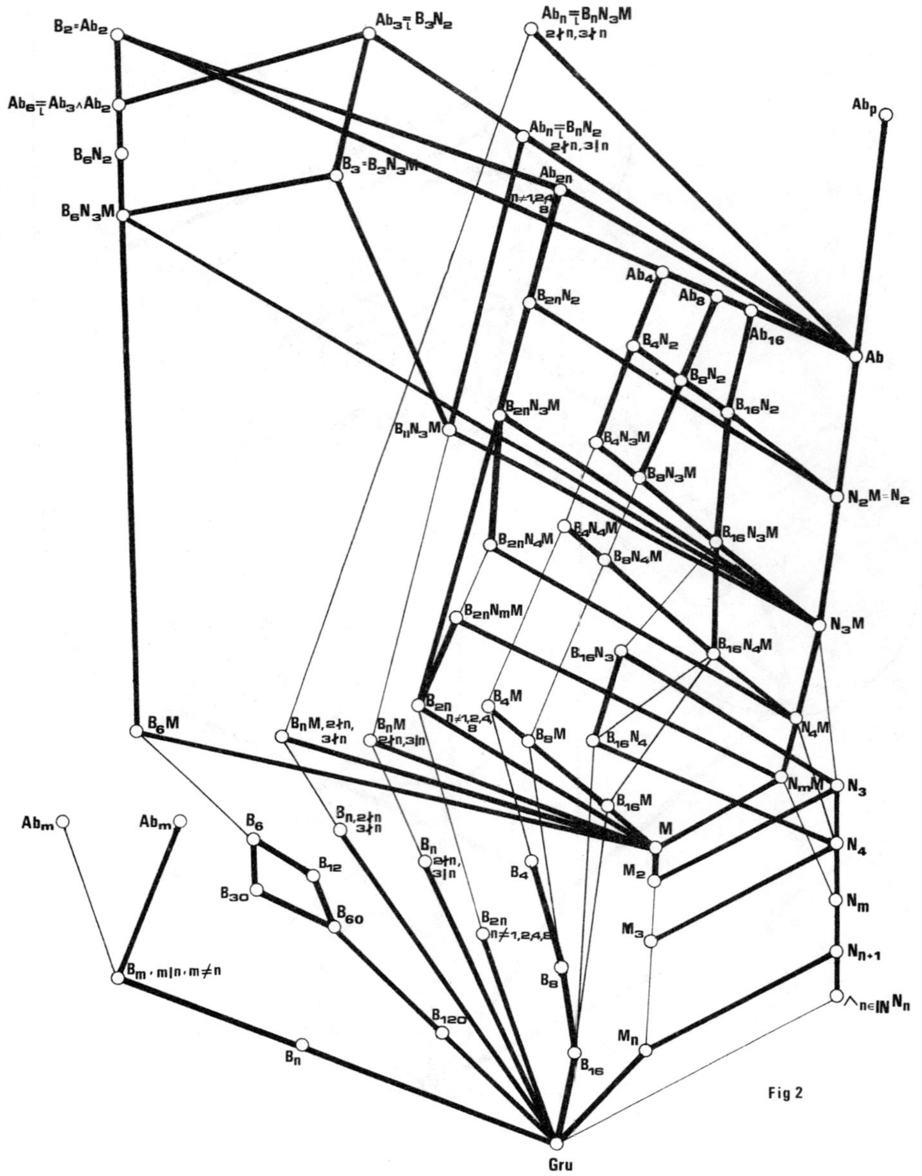

$B_2 = Ab_2$
$Ab_3 =_{\iota} B_3N_2$
$Ab_n =_{\iota} B_nN_3M$
$2\nmid n, 3\nmid n$
Ab_p
$Ab_6 =_{\iota} Ab_3 \wedge Ab_2$
B_6N_2
$Ab_n =_{\iota} B_nN_2$
$2\nmid n, 3|n$
$B_3 = B_3N_3M$
Ab_{2n}
$n\neq 1,2,4,8$
B_6N_3M
Ab_4
Ab_8
Ab_{16}
Ab
$B_{2n}N_2$
B_4N_2
B_8N_2
$B_{16}N_2$
$B_{2n}N_3M$
$B_{11}N_3M$
B_4N_3M
B_8N_3M
$N_2M = N_2$
$B_{2n}N_4M$
B_4N_4M
$B_{16}N_3M$
B_8N_4M
$B_{2n}N_mM$
N_3M
$B_{16}N_3$
$B_{16}N_4M$
B_{2n}
$n\neq 1,2,4,8$
B_4M
B_6M
$B_nM, 2\nmid n, 3\nmid n$
B_nM
$2\nmid n, 3|n$
B_8M
$B_{16}N_4$
N_4M
N_mM
N_3
$B_{16}M$
M
N_4
Ab_m
Ab_m
B_6
$B_n, 2\nmid n, 3\nmid n$
B_{12}
B_{30}
B_{60}
B_n
$2\nmid n, 3|n$
B_4
M_2
N_m
M_3
B_{2n}
$n\neq 1,2,4,8$
N_{n+1}
$B_m, m|n, m\neq n$
B_8
$\bigwedge_{n\in\mathbb{N}} N_n$
B_{120}
M_n
B_n
B_{16}
Gru

Fig 2

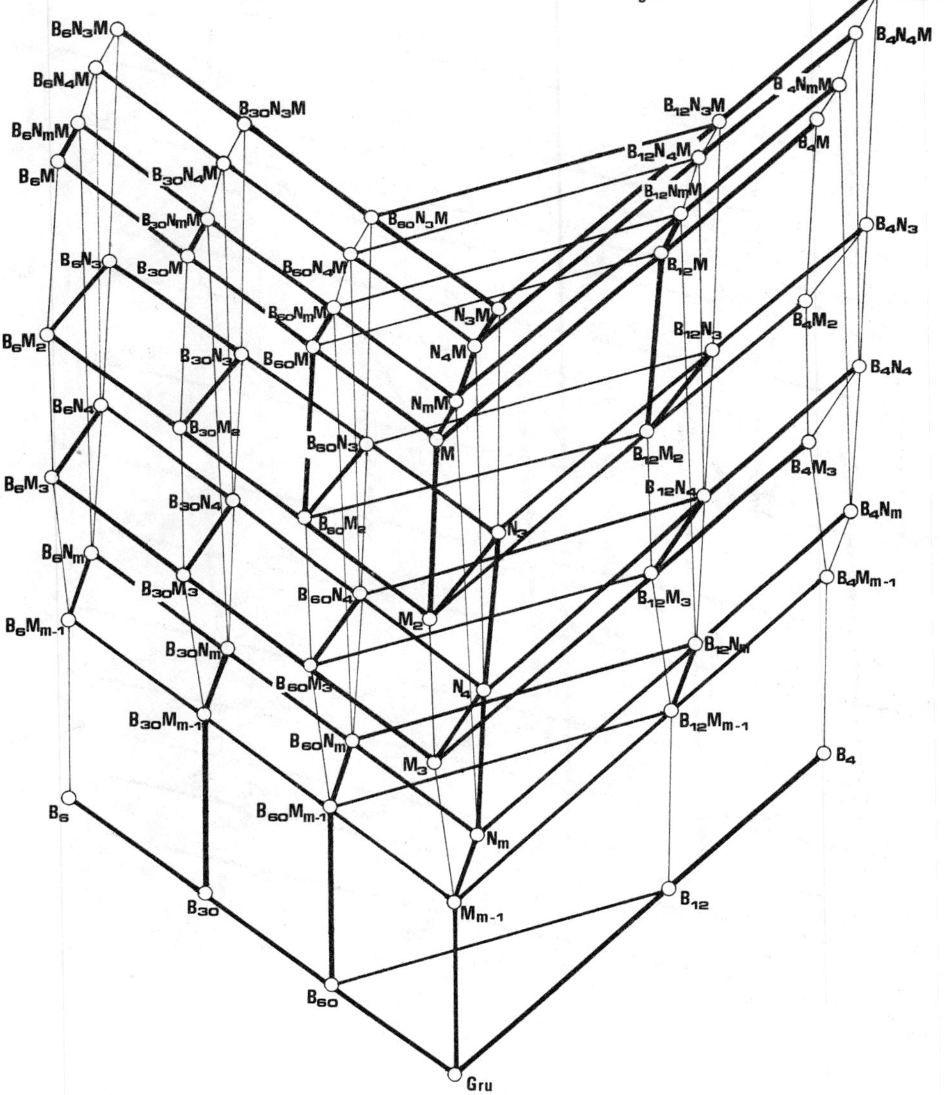
B6N3M
B6N4M
B6NmM
B6M
B6N3
B6M2
B6N4
B6M3
B6Nm
B6Mm-1
B6
B30N3M
B30N4M
B30NmM
B30M
B30N3
B30M2
B30N4
B30M3
B30Nm
B30Mm-1
B30
B60N3M
B60N4M
B60NmM
B60M
B60N3
B60M2
B60N4
B60M3
B60Nm
B60Mm-1
B60
N3M
N4M
NmM
M
N3
M2
N4
M3
Nm
Mm-1
Gru
B12N3M
B12N4M
B12NmM
B12M
B12N3
B12M2
B12N4
B12M3
B12Nm
B12Mm-1
B12
B4N3M
B4N4M
B4NmM
B4M
B4N3
B4M2
B4N4
B4M3
B4Nm
B4Mm-1
B4

Fig 3

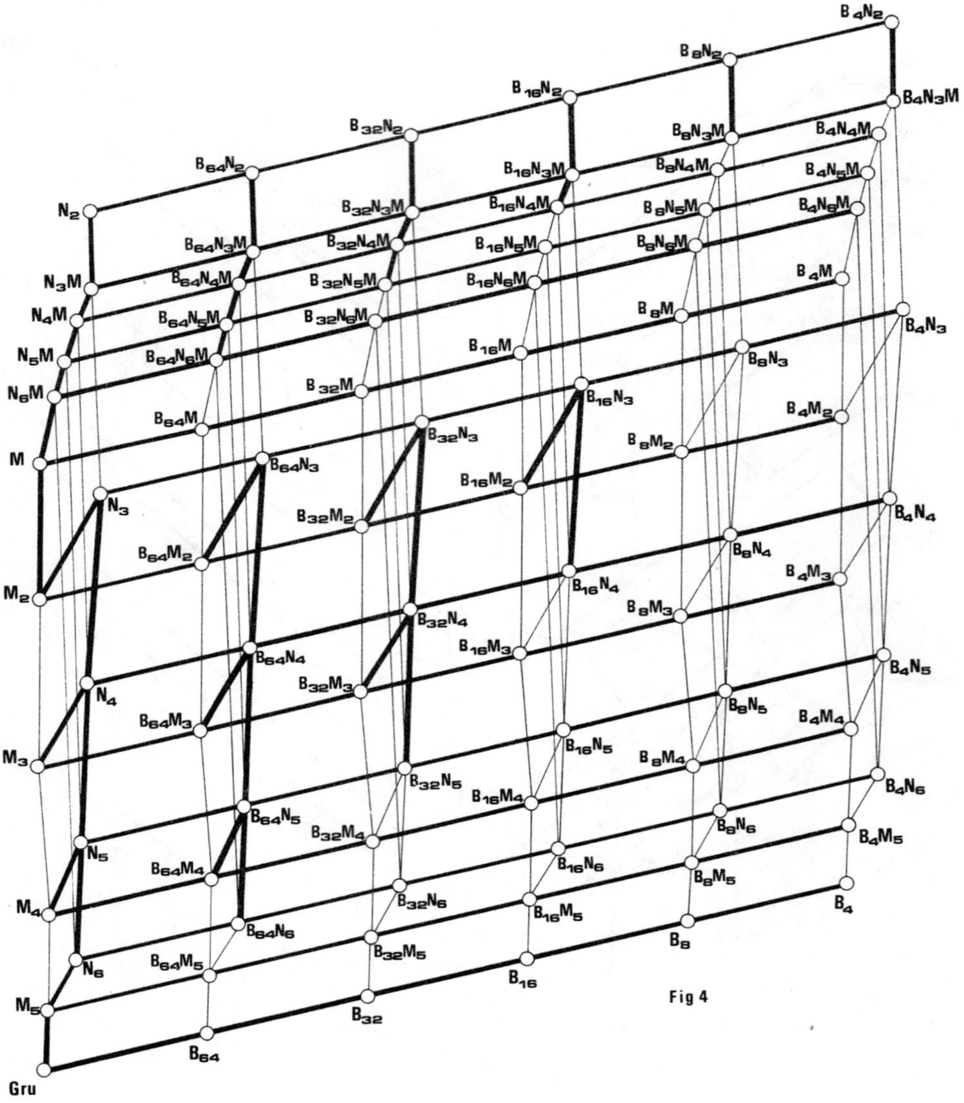
B_4N_2
B_8N_2
$B_{16}N_2$
$B_{32}N_2$
$B_{64}N_2$
N_2
B_4N_3M
B_8N_3M
B_4N_4M
$B_{16}N_3M$
B_8N_4M
B_4N_5M
$B_{32}N_3M$
$B_{16}N_4M$
B_8N_5M
B_4N_6M
$B_{64}N_3M$
$B_{32}N_4M$
$B_{16}N_5M$
B_8N_6M
N_3M
$B_{64}N_4M$
$B_{32}N_5M$
$B_{16}N_6M$
B_4M
N_4M
$B_{64}N_5M$
$B_{32}N_6M$
B_8M
B_4N_3
N_5M
$B_{64}N_6M$
$B_{16}M$
B_8N_3
N_6M
$B_{32}M$
$B_{16}N_3$
$B_{64}M$
$B_{32}N_3$
B_4M_2
M
$B_{64}N_3$
B_8M_2
$B_{16}M_2$
N_3
$B_{32}M_2$
B_4N_4
$B_{64}M_2$
B_8N_4
M_2
$B_{16}N_4$
B_4M_3
B_8M_3
$B_{32}N_4$
$B_{16}M_3$
$B_{64}N_4$
B_4N_5
N_4
$B_{32}M_3$
$B_{64}M_3$
B_8N_5
B_4M_4
$B_{16}N_5$
M_3
B_8M_4
$B_{32}N_5$
B_4N_6
$B_{16}M_4$
$B_{64}N_5$
B_8N_6
$B_{32}M_4$
B_4M_5
N_5
$B_{16}N_6$
$B_{64}M_4$
B_8M_5
M_4
$B_{32}N_6$
B_4
$B_{16}M_5$
$B_{64}N_6$
B_8
$B_{32}M_5$
$B_{64}M_5$
N_6
B_{16}
M_5
B_{32}
B_{64}
Gru

Fig 4

Fig 5

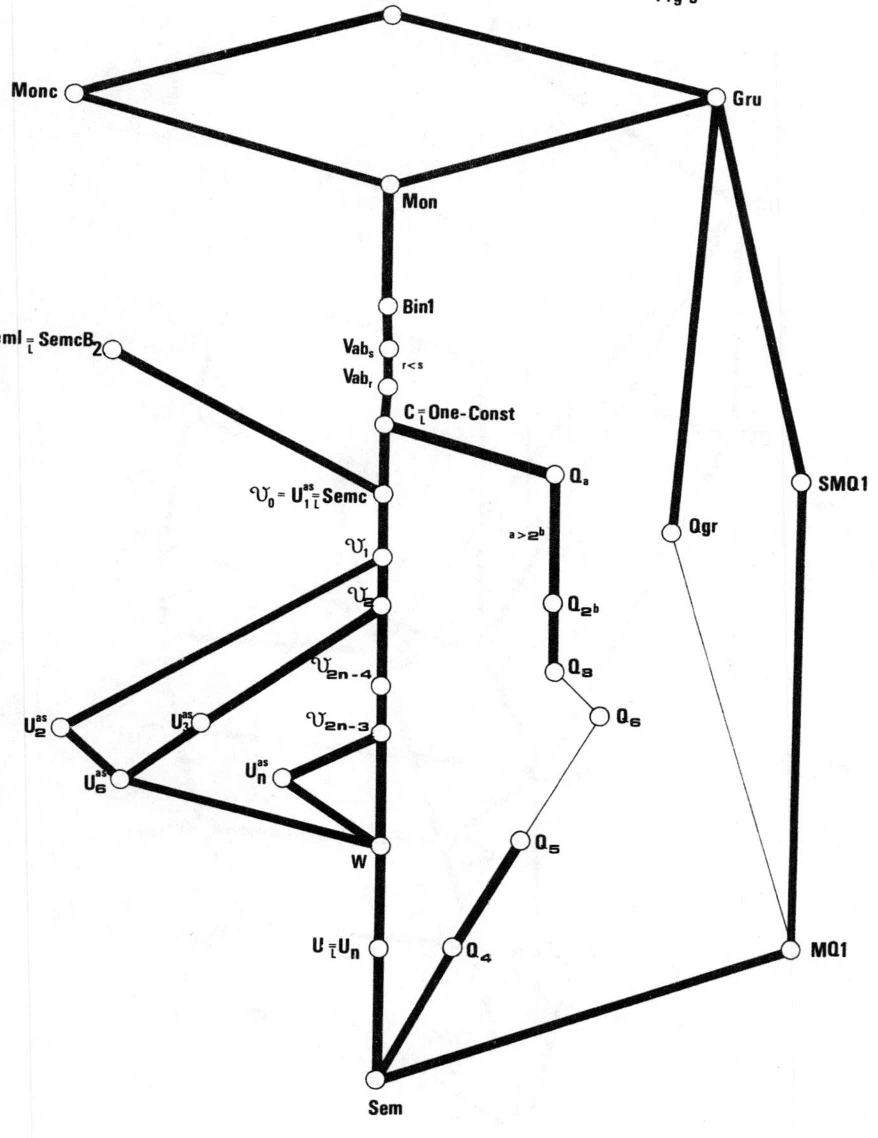
Ab
Monc
Gru
Mon
Bin1
Seml = SemcB2
Vabs
r<s
Vabr
C = One-Const
Qa
SMQ1
V0 = U1as = Semc
Qgr
a>2b
V1
V2
Q2b
V2n-4
Q3
U2as
U3as
V2n-3
Q6
U6as
Unas
W
Q5
U = Un
Q4
MQ1
Sem

Fig 6

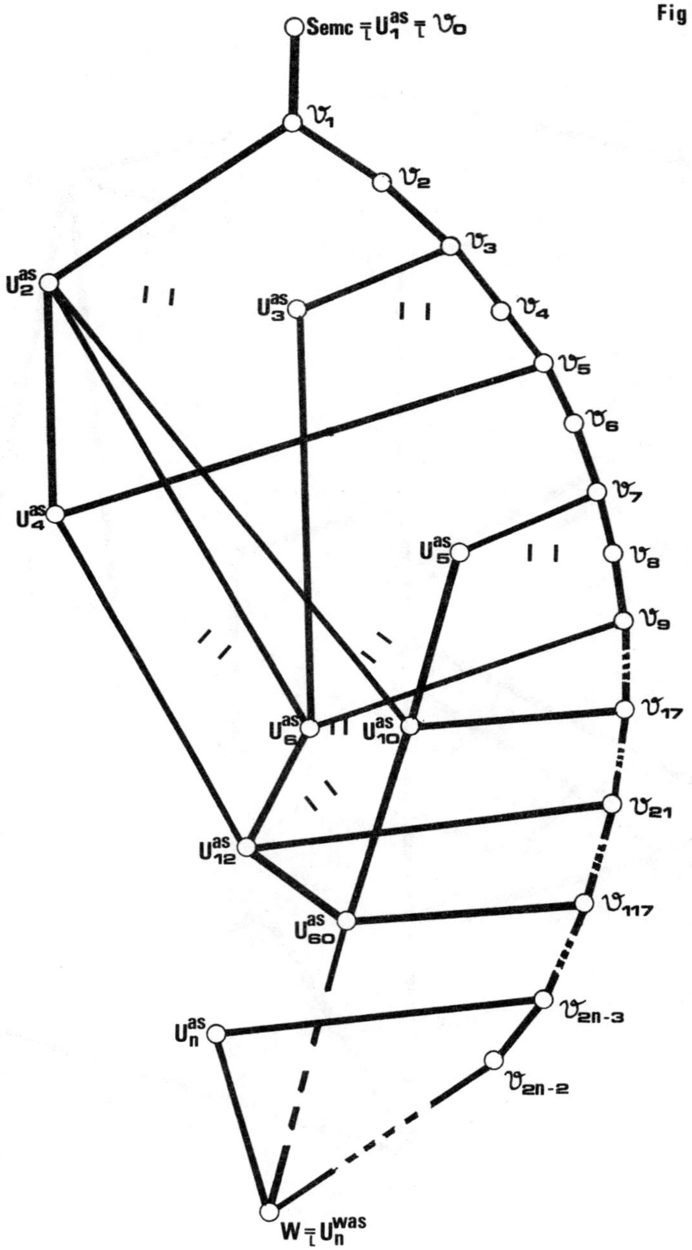
Semc ≡ U₁^as ≡ ϑ₀
ϑ₁
ϑ₂
ϑ₃
ϑ₄
ϑ₅
ϑ₆
ϑ₇
ϑ₈
ϑ₉
ϑ₁₇
ϑ₂₁
ϑ₁₁₇
ϑ₂ₙ₋₃
ϑ₂ₙ₋₂
U₂^as
U₃^as
U₄^as
U₅^as
U₆^as
U₁₀^as
U₁₂^as
U₆₀^as
Uₙ^as
W ≡ Uₙ^was

Fig 7

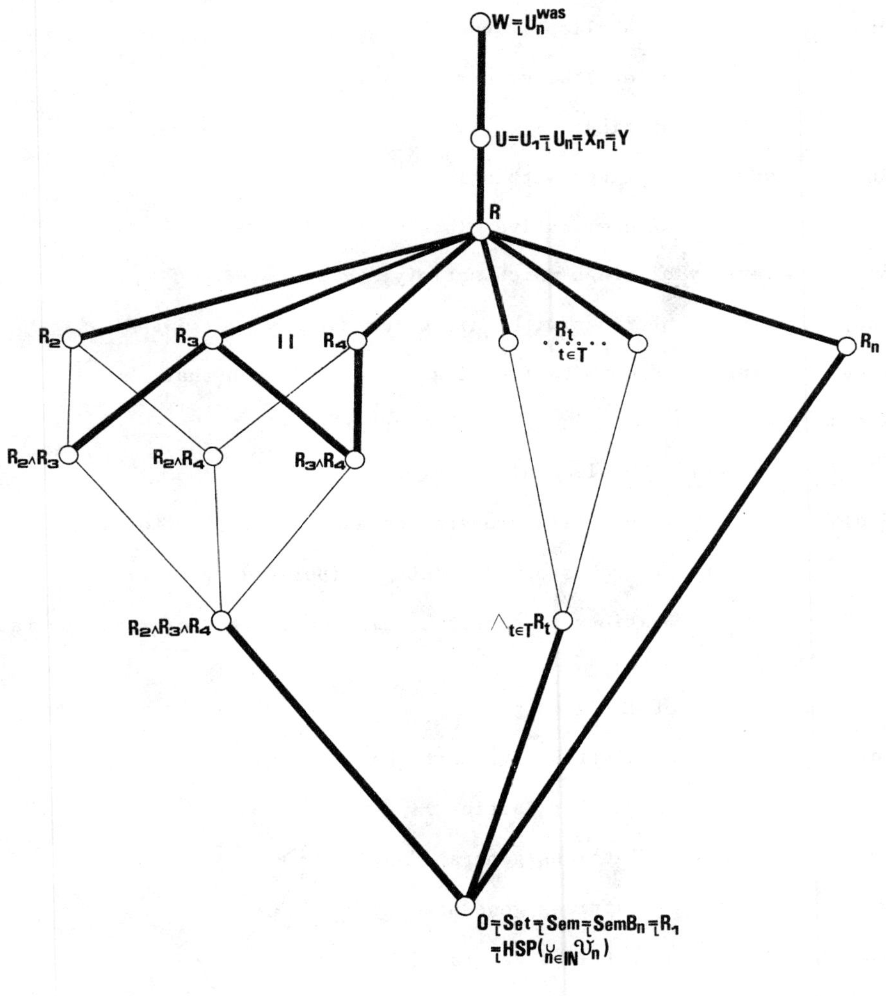
W ⊑ U^was_n
U = U_1 ⊑ U_n ⊑ X_n ⊑ Y
R
R_2
R_3
||
R_4
R_t
t∈T
R_n
R_2∧R_3
R_2∧R_4
R_3∧R_4
R_2∧R_3∧R_4
∧_t∈T R_t
O ⊑ Set ⊑ Sem ⊑ SemB_n ⊑ R_1
⊑ HSP(∪_n∈ℕ 𝔙_n)

Notation Used in the Figures

(This notation will be followed as much as possible throughout the memoir, especially in §§9-11.)

Ab variety of abelian groups

Ab_n variety of abelian groups of exponent n

An variety of rings

Anu variety of rings with unit

Ancu variety of commutative rings with unit

An_k variety of rings which satisfy $kx \approx 0$

Anu_k variety of rings with unit which satisfy $k \cdot 1 \approx 0$

$Ancu_k$ variety of commutative rings with unit such that $k \cdot 1 \approx 0$

Ancup subvariety of $Ancu_p$ such that $x^p \approx x$

BA variety of boolean algebras

Bin1 variety with a binary operation and an identity for it

B_n variety of groups of exponent n (Burnside variety)

C (One-Constant) variety of constant semigroups, or a constant operation of any arity

Lat variety of lattices

Ld variety of distributive lattices

Lm variety of modular lattices

Lm_5 variety of lattices generated by 

Lm_6 variety of lattices generated by 

Ln_5 variety of lattices generated by 

M variety of metabelian groups

M_n variety of n-metabelian groups. See page 84 .

Mt_0 variety of majority equations $mxxy \approx mxyx \approx myxx \approx x$

Mt_1 variety of Mal'tsev equations $pxyy \approx pyyx \approx x$

Mt_2 variety of equations $pxxy \approx pxyx \approx pyxx \approx y$

Mt_3	variety of equations $mxxy \approx myxx \approx x$
Mon	variety of monoids
Monc	variety of commutative monoids
MQ1	variety with three binary operations $(\cdot, \backslash, /)$ such that $x \backslash (x \cdot y) \approx y$ and $(x \cdot y)/y \approx x$
N_n	variety of n-nilpotent groups. See page 83 .
$\mathcal{N}_k$	variety of vector spaces over a field k of characteristic 0
Qgr	variety of quasigroups: three binary operations $(\cdot, \backslash, /)$ and Q1 : $x \backslash (x \cdot y) \approx y$, $(x \cdot y)/y \approx x$ Q2 : $x \cdot (x \backslash y) \approx y$, $(x/y) \cdot y \approx x$
Q_k	variety introduced in the proof of Proposition 83
R	variety with one binary operation satisfying for all n : $(x_1 \ldots x_n)(x_n \ldots x_1 y) \approx (x_n \ldots x_1 y)(x_1 \ldots x_n)$
R_n	variety with one binary operation satisfying for n : $(x_1 \ldots x_n)(x_n \ldots x_1 y) \approx (x_n \ldots x_1 y)(x_1 \ldots x_n)$
Sem	variety of semigroups
$SemB_n$	variety of semigroups satisfying $x^n \approx x$
Semc	variety of commutative semigroups
$Sem\ell$	variety of semilattices
SMQ1	variety with three binary operations $(\cdot, \backslash, /)$ such that $\cdot$ is associative and Q1 (above) is satisfied.
$\mathcal{U}$	variety with one binary operation satisfying $xy \approx yx$
$\mathcal{U}_n$	variety with one binary operation satisfying $x^n y^n \approx y^n x^n$
$\mathcal{U}_n^{as}$	subvariety of $\mathcal{U}_n$ where $(xy)z \approx x(yz)$
$\mathcal{U}_n^{was}$	subvariety of $\mathcal{U}_n$ where $(xy)(uv) \approx (xu)(yv)$
$\mathcal{V}_n$	subvariety of Sem where $yx_1 \ldots x_n z \approx zx_1 \ldots x_n y$
V_n	variety of Post algebras discussed in § 2
Vab_r	a class of varieties defined in § 11
V(•—•)	this is the variety V(A) defined at the beginning of § 2, with A taken to be a closed interval
V(O)	similar, with O denoting a circle (i.e., S_1)
W	subvariety of $\mathcal{U}$ where $(xy)(uv) = (xu)(yv)$

X_n variety with one binary operation satisfying:

$$x_1 \dots x_{n-1}x_n \approx x_n(x_1 \dots x_{n-1})$$

Y variety with one binary operation satisfying:

$$x_1x_2x_3 \approx x_3(x_1x_2) \quad \text{and} \quad x_1x_2x_3 \approx x_2x_1x_3$$

REMARK. When we write a product $a_1a_2 \dots a_n$ involving a binary operation, we mean $(\dots (((a_1a_2)\cdot a_3)\cdot a_4) \dots)a_n$.

1. PRELIMINARIES

In the introduction we stated one definition of the relation $V \leq W$ for varieties V and W. It will be useful for us to consider an equivalent condition. Let us think of the class of all homomorphisms $f: \mathfrak{A} \to \mathfrak{B}$ (with $\mathfrak{A}, \mathfrak{B} \in V$) as a *category*, which we will also denote V. (Its morphisms are all homomorphisms between non-empty algebras in V.) As a category, V has an underlying set functor $U_V : V \to \text{Sets}$ (also called the forgetful functor to sets, or the universe functor). It is easy to see that the correspondence $(A\ ;\ G_s)_{s\in S} \longrightarrow (A\ ;\ \overline{\alpha}_t)_{t\in T}$, mentioned in the introduction, is in fact a functor $\Phi : W \to V$, which commutes with the underlying set functors, i.e.:

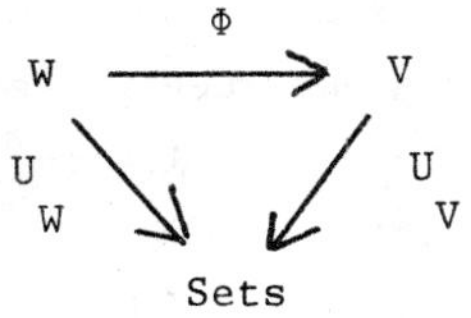

is a commutative diagram. Thus we have established one half of the equivalence in the following Proposition. We omit the proof of the other half (the converse of what we have just proved), since it is well known and is in any case not used in our work here. A functor Φ obeying the following conditions is called a *concrete functor.*

Proposition 1. *$V \leq W$ iff there exists a functor $W \overset{\Phi}{\to} V$ which commutes with the underlying set functor.* ∎

(For a detailed proof see [21] or [53].)

There are two especially simple types of concrete functor Φ. First, if V is a *reduct* of W, i.e., if V is formed by omitting some of the operations of W (and then closing under H, S and P), then Φ is simply the "forgetful functor" which omits the appropriate operations. Second, if W is a *subvariety* of V, then Φ is simply the inclusion map. It can be proved that (within equivalence) every concrete functor can be factored as $\Phi = \Psi\Theta$ with Θ a reduct map and Ψ an inclusion map.

Varieties V and W are called *equivalent* iff there exists a *bijective* functor Φ of the type described in Proposition 1. (For some equivalent conditions, see [53], [72] or [74].) We will need, in this paper, to work with a coarser equivalence relation on the class of all varieties, since we wish the order relation $\leq$ to be anti-symmetric (it is clearly reflexive and transitive). It is not difficult to observe that there exist inequivalent varieties V and W with $V \leq W$ and $W \leq V$, for example V = sets (algebras with no operations), W = semigroups. (For $W \leq V$ define a semigroup operation xy in the language of sets *via* xy = x.) To remedy this situation, we will write $V \equiv W$ for $V \leq W$ and $W \leq V$ ("V is equi-interpretable with V") and consider the equivalence classes under $\equiv$. As is well known, these classes are partially ordered if we extend $\leq$ to equivalence classes in the obvious way: $[V] \leq [W]$ means $V \leq W$. Here $[V]$ means the $\equiv$-equivalence class of V; sometimes, as in our diagrams, we will allow ourselves to omit the brackets when no confusion will result. We will also use $[\Sigma]$ to denote $[V]$ for V the variety defined by a set Σ of identities. Thus we have now created a partially ordered class

$$L = (\{[V] : V \text{ any variety}\}, \leq).$$

The reader may have been puzzled by our remark that semigroups $\leq$ sets, since the variety of semigroups seems much richer than that of sets. But this paradox is resolved by thinking of the relation $V \leq W$ as meaning, loosely, that it is at least as hard to build one model of W as it is to build one model of V (for any time we construct $\mathfrak{A} \in W$, we have really already constructed $\Phi\mathfrak{A}$, which is in V). Then it is clear that from some other points of view, V may be the "stronger" or "richer" variety, since in most cases we will have $\mathfrak{B} \in V$ not of the form $\Phi\mathfrak{A}$. Even so, it has proved useful to sometimes think of W as stronger than V, since many important varietal properties which hold for V will also hold for W, such as modularity of congruence lattices of algebras in V. This observation is the basis of the theory of Mal'tsev conditions, which of course was the starting point for the theory of L (see §5 below for a brief recapitulation of the theory of Mal'tsev conditions).

Proposition 2. L *is a complete lattice, i.e., if* K *is any subclass of* L *which is a set, then* K *has a greatest lower bound and a least upper bound in* L.

We have stated this so fussily because it will turn out that L is indeed a proper class. (Even so, we do not yet know any example of a subclass of L with no infimum or with no supremum.) Notice that we cannot prove Proposition 2 by the usual device of, say, first proving that suprema exist, and then saying that inf K = sup K' for K' the set of lower bounds of K; here this will not work because K' will generally be a proper class. Rather than repeat the proof given in [63], we will present a new proof of this result which we learned from Jan Mycielski. We thank Professor Mycielski for permitting us to include his proof, which actually allows us to retain the above-mentioned construction of inf K as sup K'. The Mycielski construction of the infimum is very different from the original one of W. D. Neumann, who actually constructed a coproduct in the category of varieties and concrete-functors.

Proof of Proposition 2. (J. Mycielski) It is clear that every set K of varieties has a supremum: we form the theory generated by the "disjoint union" of the theories of all varieties in K. (For more details, see the proof of Proposition 3.)

To construct the infimum or meet, $\bigwedge K$, we first form the class K' of all lower bounds. As usual, it will suffice to prove that K' has a supremum. And for this, it will suffice to show the existence of a cofinal subset of K', i.e., a *set* $K'' \subseteq K'$ such that for all $V \in K'$ there exists $W \in K''$ with $V \leq W$. Now the desired set K" may be very simply defined: it is the subset of K' consisting of all (equivalence classes of) varieties with $\leq \kappa$ operations, where

$$\kappa = \Pi\{\kappa(U) : U \in K\}$$
$$\kappa(U) = \aleph_0 + \text{the number of operations of } U.$$

(The existence of this product of cardinals is where we use the fact that K itself is a set.)

Now it remains only to show that K" has the required cofinality property: we will be given $V \in K'$ and find $W \in K''$ with $W \geq V$. Since V is a lower bound of K, we have, for each $U \in K$, an interpretation

$$F_t \to \alpha_t^U$$

where F_t ranges over all V-operations, and α_t^U is a term in the language of U. Define an equivalence relation θ on the set $\{F_t\}$ of all V-operations by setting

$$F_t \theta F_s \leftrightarrow (\forall U \in K)(\alpha_t^U = \alpha_s^U)$$

(where the last equation means that these two terms are formally identical). We may now define W to be the subvariety of V defined by the identities

$$F_t x_1 x_2 \ldots \approx F_s x_1 x_2 \ldots$$

for all (t,s) with $F_t \theta F_s$. It is obvious that $V \leq W$. To see that W remains a lower bound of K, we retain, for each $U \in K$, the same interpretation of V in U as we had above; the definition of θ tells us that this interpretation also interprets the new identities of W. Thus W is a lower bound of K, i.e., $W \in K'$. Now elementary cardinal arithmetic tells us that (within equivalence) W has at most κ operations, and so $W \in K''$. Thus we have established the cofinality of K'' in K'. ∎

The existence assertions of the next Proposition are certainly redundant, but we need the useful detailed description of $[V] \wedge [W]$. The ideas involved are certainly not new (see references given in [74; pp. 265-268]), but perhaps this presentation is a little different from what has come before. For the next few pages we will be assuming that V and W have disjoint similarity types, i.e., that the V-operations are $F_s (s \in S)$ and the W-operations are $F_s (s \in T)$, with $S \cap T = \emptyset$. Moreover, we will assume that $\Sigma = \{\sigma_i \approx \tau_i : i \in I\}$ is an equational axiom base for V, and likewise $\Gamma = \{\mu_j \approx \nu_j : j \in J\}$ for W.

Definition. The *coproduct* variety $V \amalg W$ is the variety of type $S \cup T$ axiomatized by $\Sigma \cup \Gamma$. The *product* variety $V \times W$ has operations $F_s (s \in S \cup T)$ and one further binary operation denoted simply by juxtaposition. Its defining identities are:

$$xx \approx x$$
$$(xy)(uv) \approx xv$$
$$F_s(x_1y_1, x_2y_2, \ldots) \approx (F_s x_1 x_2 \ldots)(F_s y_1 y_2 \ldots) \qquad (s \in S \cup T)$$
$$(F_s x_1 x_2 \ldots) \cdot y \approx x_1 y \qquad (s \in T)$$
$$x \cdot (F_s y_1 y_2 \ldots) \approx x y_1 \qquad (s \in S)$$
$$\sigma_i \cdot y \approx \tau_i \cdot y \qquad (i \in I)$$
$$x \cdot \mu_j \approx x \cdot \nu_j \qquad (j \in J)$$

Proposition 3. *$[V] \vee [W]$ exists and equals $V \amalg W$, and $[V] \wedge [W]$ exists and equals $V \times W$.*

Proof. One may easily check that $V \amalg W$ is a least upper bound directly from the definition of interpretability; we omit the details. For $[V] \wedge [W]$, we first observe that $V \times W$ can easily be interpreted in V by defining xy to be x, F_s to be F_s for $s \in S$, and $F_s(x_1, x_2, \ldots)$ to be x_1 for $s \in T$. Thus $V \times W \leq V$, and similarly $V \times W \leq W$. Now suppose that U is any other lower bound of V and W. This means that for the U-operations G_t there are V-terms α_t interpreting U in V, and likewise W-terms β_t interpreting U in W. We omit the straightforward inductive proof that the $V \times W$-terms $\alpha_t \beta_t$ will interpret U in $V \times W$. Alternatively, one may see that $U \leq V \times W$ by using Proposition 1 together with Proposition 4 below, as follows. There are functors $\Phi : V \to U$ and $\Psi : W \to U$

commuting with the underlying set functor. To obtain such a functor $\Theta : V \times W \to U$, simply factor each homomorphism in $V \times W$ as $\alpha \times \beta : \mathfrak{A} \times \mathfrak{B} \to \mathfrak{C} \times \mathfrak{D}$ ($\mathfrak{A}, \mathfrak{C} \in V$, $\mathfrak{B}, \mathfrak{D} \in W$), and define $\Theta(\alpha \times \beta) = \Phi(\alpha) \times \Psi(\beta)$. ∎

For simplicity in stating the next proposition and in the remainder of the paper, we will assume that V *is* the subvariety of $V \times W$ defined by the additional law $xy \approx x$, and that W *is* the subvariety of $V \times W$ defined by $xy \approx y$. [These subvarieties are clearly equivalent to the given V and W; in fact, they differ only by the addition of some extra trivial operations.] This Proposition was known to F. W. Lawvere at least by 1968.

Proposition 4. *Every algebra in* $V \times W$ *is isomorphic to* $\mathfrak{A} \times \mathfrak{B}$ *for some* $\mathfrak{A} \in V$, $\mathfrak{B} \in W$. *Every homomorphism* $\gamma : \mathfrak{A}_1 \times \mathfrak{B}_1 \to \mathfrak{A}_2 \times \mathfrak{B}_2$ ($\mathfrak{A}_i \in V$, $\mathfrak{B}_i \in W$) *has the form* $\gamma(a,b) = (\alpha(a), \beta(b))$ *for some unique homomorphisms* $\alpha : \mathfrak{A}_1 \to \mathfrak{A}_2$, $\beta : \mathfrak{B}_1 \to \mathfrak{B}_2$. Such γ may be denoted $\alpha \times \beta$.

Proof. For $\mathfrak{C}$ any algebra in $V \times W$, choose $b \in C$ and let Cb denote $\{cb : c \in C\}$, and likewise bC. Define maps

$$C \underset{\psi}{\overset{\phi}{\rightleftarrows}} Cb \times bC$$

via $\phi(x) = (xb, bx)$ and $\phi(xb, by) = xy$. One easily verifies, from the first two defining identities of $V \times W$, that $\psi \circ \phi$ and $\phi \circ \psi$ are both identity maps, i.e., that ψ and ϕ are bijections and that $\psi = \phi^{-1}$. For each $V \times W$-operation F_s, define an operation F'_s on Cb and an operation F''_s on bC as follows:

$$F'_s(x_1b, x_2b, \ldots) = F_s(x_1b, x_2b, \ldots) \cdot b$$
$$F''_s(bx_1, bx_2, \ldots) = b \cdot F_s(bx_1, bx_2, \ldots)$$

Let $\mathfrak{A} = (Cb; F'_s)_{s\in S}$ and $\mathfrak{B} = (bC; F''_s)_{s\in S}$. One easily checks that ϕ is a homomorphism, that $\mathfrak{A}$ obeys the identity $xy \approx x$ and that $\mathfrak{B}$ obeys $xy \approx y$. Thus we have completed our factorization of $\mathfrak{C}$. The remaining assertion about γ is quite easy, and we omit the proof. ∎

The method of the last proof actually extends much farther. Consult Kelly and Pultr [47] for general category-theoretic conditions yielding such results. We will only look at what happens for a *topological algebra* in $V \times W$. (This means a structure $A = (\mathfrak{A}; T)$ with $\mathfrak{A}$ an algebra and T a topology on the universe A of $\mathfrak{A}$ such that each operation $F : A^n \to A$ is continuous. If $\mathfrak{A} \in V$, we will say that A is a "topological algebra in V."

Proposition 5. *Every topological algebra in* $V \times W$ *is isomorphic to a product* $A \times B$ *with* A *a topological algebra in* V *and* B *a topological algebra in* W.

Proof. (A continuation of the proof of Proposition 4.) Let us suppose that $C = (\mathbf{C}, T)$ is a topological algebra in $V \times W$. The algebras $\mathfrak{A}$ and $\mathfrak{B}$ of the former proof have universes Cb and bC, which are subsets of the topological space (C,T). Now if we give Cb and bC their subspace topologies, it is clear from the defining formulas that the operations F_s' and F_s'' are continuous, making $\mathfrak{A}$ and $\mathfrak{B}$ into topological algebras. Likewise ϕ and ψ are continuous, yielding the required factorization of C as a topological algebra. ∎

We emphasize that it is possible to have $[U] = [V] \vee [W]$ or $[U] = [V] \wedge [W]$ in our lattice without U being equivalent to $V' \amalg W'$ or $V' \times W'$ for any V' and W'. Some examples for $\vee$ will be given in §5 (after Prop. 32) and §6 (U defined by the majority identities). We do not know any naturally occurring example for $\wedge$; the variety of Newmann algebras, described in pages 49-51 of Birkhoff [7], while not axiomatized like our $V \times W$, is equivalent to such a $V \times W$. In fact, it will be our contention later on in §4 that almost all naturally occuring varieties are $\wedge$-irreducible in L.

For our examples we will find it convenient to have a construction of the "k-th power variety" $V^{[k]}$ of a given variety V, which has the significant property that the models of $V^{[k]}$ are precisely the kth powers of the models of V (in a sense made precise below). For a full theory of this construction, see either [56] or pp. 268-269 of Taylor [74]; here we will present only what we need. In the next definition, V will be presented by Σ exactly as it was before the last definition. Here k denotes a finite integer.

Definition. The kth power variety $V^{[k]}$ has operations F_s $(s \in S)$, a new k-ary operation d, and a new unary operation c. Its defining identities are:

$$d(x, \ldots, x) \approx x$$
$$d(d(x_{11}, \ldots, x_{1k}), \ldots, d(x_{k1}, \ldots, x_{kk})) \approx d(x_{11}, \ldots, x_{kk})$$
$$d(cx_1, \ldots, cx_k) \approx cd(x_2, \ldots, x_k, x_1)$$
$$c^k x \approx x$$
$$F_s(d(x_{11}, \ldots, x_{1k}), \; d(x_{21}, \ldots, x_{2k}), \; \ldots)$$
$$\approx d(F_s(x_{11}, x_{21}, \ldots), \; \ldots, \; F_s(x_{1k}, x_{2k}, \ldots)) \quad (s \in S)$$
$$\sigma_i \approx \tau_i \quad (i \in I).$$

We prove the next proposition only for $k = 2$; for a complete proof of this and related results, see Taylor [loc. cit.]. B. A. Davey and H. Werner have shown on p. 197 of [17] that no such result is possible for infinite k; see also [3].

Proposition 6. *If* $\mathbb{L} \in V^{[k]}$, *then the* V*-reduct of* $\mathbb{L}$ *is isomorphic to* $\mathfrak{A}^k$ *for some* $\mathfrak{A} \in V$. *If* C *is a topological algebra in* $V^{[k]}$ *then the* V*-reduct of* C *is isomorphic to* A^k *for some topological algebra* A *in* V.

Proof. (k = 2 only. We denote d(x,y) by xy and then continue the proofs of Propositions 4 and 5.) Every model of $V^{[2]}$ is of course a model of $V \times V$; the only thing we need to add to the previous proof is to show the existence of an isomorphism $\lambda : \mathfrak{A} \cong \mathfrak{B}$ which is bicontinuous in the topological case. First note that we may assume that b is fixed under c, i.e., c(b) = b. For, if not, we replace b by $b' = b \cdot c(b)$; then by our identities $c(b') = c(b \cdot c(b)) = c^2(b) \cdot c(b) = b \cdot c(b) = b'$. Then we may simply define $\lambda : Cb \to bC$ *via* $\lambda = c$. It is almost immediate that this λ is a (bicontinuous) isomorphism (with $\lambda^{-1} = c$).∎

As an example for the next proposition, we will use the variety U axiomatized by these identities:

$$g_1f(x,y) \approx x \qquad g_2f(x,y) \approx y$$
$$f(g_1x, g_2x) \approx x.$$

In any model $(A;f,g_1,g_2)$ the operation $f : A^2 \to A$ is obviously bijective; thus A has no finite models other than singletons, and if A is topological model of U, then $A \cong A^2$ (as a space). (U and some related varieties will be discussed again later in §2 after Proposition 11).

Proposition 7. *If* $V \leq W$, *then* $V^{[k]} \leq W^{[k]}$, *and so* $[V] \to [V^{[k]}]$ *is well defined as a map on* L; *moreover it is a* $\wedge$*-semilattice homomorphism.* $V \leq V^{[k]}$, *and if* V *has a non-trivial finite model (or a topological model of finite dimension* > 0*), then* $[V] < [V^{[k]}]$ *for* $k \geq 2$. *There exists a non-trivial variety* U *with* $[U] = [U^{[2]}]$.

Proof. The first sentence is more or less obvious. For $V \leq V^{[k]}$ we note that V is evidently a reduct of $V^{[k]}$. If n is the smallest cardinality of any non-trivial finite model in V, then $V^{[k]}$ has no model of power n (its smallest model being of power n^k). Thus there cannot be any universe-preserving functor $V^{[k]} \to V$, and so $[V] < [V^{[k]}]$. A similar argument applies to topological dimension (and any of the various dimension theories will be adequate for this purpose; see e.g. [61]). Finally, we will show that $U^{[2]} \leq U$ for U the variety given just before the theorem. $U^{[2]}$ may be axiomatized as follows:

$$\psi_1\phi xy \approx x \qquad \psi_2\phi xy \approx y \tag{1}$$
$$\phi\psi_1x\psi_2x \approx x \tag{2}$$

$$\bar{\bar{x}} \approx x \tag{3}$$

$$\overline{xy} \approx \bar{y}\,\bar{x} \tag{4}$$

$$xx \approx x \tag{5}$$

$$(xy)(uv) \approx xv \tag{6}$$

$$\phi(x,y) \cdot \phi(u,v) \approx \phi(xu,yv) \tag{7}$$

$$\psi_i xy \approx \psi_i x \cdot \psi_i y \qquad (i=1,2) \tag{8}$$

(To avoid future ambiguity, we have redesignated g_i by ψ_i and f by ϕ. Moreover, for simplicity $c(x)$ has been denoted $\bar{x}$ and $d(x,y)$ has been denoted xy.) To prove $U^{[2]} \leq U$, we define:

$$\phi(xy) \approx f(fg_1xg_1y,\ fg_2xg_2y)$$

$$\psi_i x \approx f(g_ig_1x,\ g_ig_2x)$$

$$\bar{x} \approx f(g_2x,\ g_1x)$$

$$x \cdot y \approx f(g_1x,\ g_2y).$$

Now we have to check (1)-(8) under this definition.

(1):
$$\begin{aligned} \psi_1\phi xy &\approx f[g_1g_1f(g_1xg_1y,\ fg_2xg_2y),\ g_1g_2f(fg_1xg_1y,\ fg_2xg_2y)] \\ &\approx f[g_1fg_1xg_1y,\ g_1fg_2xg_2y] \\ &\approx f[g_1x,\ g_2x] \approx x \end{aligned}$$

likewise $\psi_2\phi xy$.

(2):
$$\begin{aligned} \phi(\psi_1x,\ \psi_2x) &\approx f[f(g_1f[g_1g_1x,\ g_1g_2x],\ g_1f[g_2g_1x,\ g_2g_2x]), \\ &\qquad f(g_2f[g_1g_1x,\ g_1g_2x],\ g_2f[g_2g_1x,\ g_2g_2x])] \\ &\approx f[f(g_1g_1x,\ g_2g_1x),\ f(g_1g_2x,\ g_2g_2x)] \\ &\approx f[g_1x,\ g_2x] \approx x \end{aligned}$$

(3):
$$\begin{aligned} \bar{\bar{x}} &\approx f[g_2f(g_2x,\ g_1x),\ g_1f(g_2x,\ g_1x)] \\ &\approx f[g_1x,\ g_2x] \approx x. \end{aligned}$$

(4):
$$\begin{aligned} \overline{xy} &\approx f[g_2f(g_1x,\ g_2y),\ g_1f(g_1x,\ g_2y)] \\ &\approx f[g_2y,\ g_1x] \\ &\approx f[g_1f(g_2y,\ g_1y),\ g_2f(g_2x,\ g_1x)] \\ &\approx \bar{y}\,\bar{x}. \end{aligned}$$

(5):
$$x \cdot x \approx f(g_1x,\ g_2x) \approx x$$

(6): $(xy)\cdot(uv) \approx f[g_1f(g_1x, g_2y), g_2f(g_1u, g_2v)]$
$\approx f[g_1x, g_2v] \approx xv$

(7): $\phi(xy)\cdot\phi(uv) \approx f[g_1f(fg_1xg_1y, fg_2xg_2y), g_2f(fg_1ug_1v, fg_2ug_2v)]$
$\approx f[fg_1xg_1y, fg_2ug_2v]$
$\approx f[f\{g_1f(g_1x, g_2u), g_1f(g_1y, g_2v)\},$
$f\{g_2f(g_1x, g_2u), g_2f(g_1y, g_2v)\}]$
$\approx \phi(xu, yv)$

(8): $\psi_i(xy) \approx f[g_ig_1f(g_1x, g_2y), g_ig_2f(g_1x, g_2y)]$
$\approx f[g_ig_1x, g_ig_2y]$
$\approx f[g_1f(g_ig_1x, g_ig_2x), g_2f(g_ig_1y, g_ig_2y)]$
$\approx \psi_i(x)\psi_i(y).$ ∎

At this point we leave the theory of $V^{[k]}$. We will return to it briefly in Proposition 11 and the remarks following, and at greater length in §7.

The above argument for $[V] < [V]^{[k]}$ employed two special cases of our general method for showing that $W \not\leq V$. Of course the method of finite algebras will often work in the situation of Proposition 7. But, for instance, the variety of *lattice-ordered groups* has no non-trivial finite models, but it does have a one-dimensional topological model (the real number system), and so Proposition 7 still yields $V^{[k]} \not\leq V$ for this variety.

The following definitions and propositions provide useful tools for obtaining $W \not\leq V$, especially in §9 and §10. With a general enough extension of the idea of "file," Proposition 8 could have "If ..., then ..." replaced by "iff."

Definition. If $\mathfrak{A} = (A; F_0F_1, F_2, \ldots)$ is any algebra, then

$$\mathrm{File}(\mathfrak{A}) = (A, \mathrm{Aut}(\mathfrak{A}), \mathrm{End}(\mathfrak{A}), \mathrm{Sub}(\mathfrak{A}), \mathrm{Con}(\mathfrak{A}), \mathrm{Top}(\mathfrak{A}), \ldots)$$

where $\mathrm{Aut}(\mathfrak{A})$ is the automorphism group of $\mathfrak{A}$, conceived of as a group of permutations of the universe A, and $\mathrm{End}(\mathfrak{A})$ is the endomorphism monoid of $(\mathfrak{A})$ (similarly conceived), $\mathrm{Sub}(\mathfrak{A})$ is its subuniverse lattice, $\mathrm{Con}(\mathfrak{A})$ is its congruence lattice, $\mathrm{Top}(\mathfrak{A})$ the lattice of all topologies making $\mathfrak{A}$ into a topological algebra, and so on. We write:

$$\mathrm{File}(\mathfrak{A}) \leq \mathrm{File}(\mathfrak{B})$$

iff $\mathfrak{A}$ and $\mathfrak{B}$ have the same universe and $\mathrm{Aut}(\mathfrak{B}) \subseteq \mathrm{Aut}(\mathfrak{A})$, $\mathrm{End}(\mathfrak{B}) \subseteq \mathrm{End}(\mathfrak{A})$, $\mathrm{Sub}(\mathfrak{B}) \subseteq \mathrm{Sub}(\mathfrak{A})$, $\mathrm{Con}(\mathfrak{B}) \subseteq \mathrm{Con}(\mathfrak{A})$, and $\mathrm{Top}(\mathfrak{B}) \subseteq \mathrm{Top}(\mathfrak{A})$, and so on.

We have included the three dots ... in our definition of File to emphasize the obvious possibility of including further such things when necessary. (See for instance §11.) The next proposition is very important to our work here, but nevertheless its proof is very easy and so we omit it.

<u>Proposition 8</u>. *If* W $\leq$ V, *then for every algebra* $\mathcal{B} \in$ V *there exists an algebra* $\mathfrak{A} \in$ W *with* File($\mathfrak{A}$) $\leq$ File($\mathcal{B}$). ■

As we already mentioned, Proposition 8 will be used in several places in this paper to show that W $\not\leq$ V for various varieties W and V. For now we content ourselves with one more example: If W is, e.g., *congruence-modular* and V is not, then W $\not\leq$ V.

In some of our applications of Proposition 8, where a "bad algebra" has to be given to ensure the falsity of W $\leq$ V, one finds a very interesting class of such bad algebras (good for our purposes), and these ideas lead to a new definition and proposition.

<u>Definition</u>. Let $\mathcal{B}$ be an algebra and U a variety. We call $\mathcal{B}$ *simple under interpretation* in U (SIN in U) iff for every two concrete functors F,G : HSP($\mathcal{B}$) $\to$ U one has F($\mathcal{B}$) $\cong$ G($\mathcal{B}$).

After the next Proposition, we provide a list of examples and non-examples of SIN algebras.

<u>Proposition 9</u>. *Let* $\mathcal{B} \in$ U *be* SIN *in* U. *If* W *and* V *are subvarieties of* U *such that* $\mathcal{B} \in$ V *but* $\mathcal{B} \notin$ W, *then* W $\not\leq$ V.

<u>Proof</u>. Suppose, by way of contradiction, that W $\leq$ V, and let Φ: V $\to$ W be a concrete functor. Thus we have two concrete functors from HSP $\mathcal{B}$ to U, namely:

$$\mathrm{HSP}\mathcal{B} \xrightarrow{i} V \xrightarrow{\Phi} W \xrightarrow{i} U \quad \text{and}$$

$$\mathrm{HSP}\mathcal{B} \xrightarrow{i} U,$$

where the three arrows labeled i are inclusion functors. The SIN property says that $\mathcal{B}$ has the same image under these two concrete functors, and so $\mathcal{B} \in$ W, which is a contradiction. ■

Examples.

1. The semigroup S defined by

	a	b
a	a	a
b	b	b

is SIN in *every variety*. If this semigroup does not belong to variety W, then we can say that $W \not\leq$ Sem (= 0). The existence of S (or its dual) in a variety V (of groupoids) is equivalent to the condition given in Proposition 10 below for a variety V to be zero. This example will be (or could be) used in the proofs of Propositions 64, 71 and 74 below.

2. The quasigroup Q defined by

	a	b	c
a	b	a	c
b	a	c	b
c	c	b	a

is SIN in the variety of groups, and helps us to obtain Qgr < Gru in Proposition 65.

3. The quasigroup H defined by

	a	b	c
a	a	c	b
b	c	b	a
c	b	a	c

is SIN in the variety of quasigroups, and also in U_n^{as} (see §10). This leads to $W < U_n^{as}$ in Proposition 80, and could also provide an alternative proof of Proposition 68.

4. As shown in Proposition 50 of §9, all cyclic groups Z_n are SIN in Gru and in C. This fact is applied for groups in the proofs of Propositions 52 and 53, as well as in some lemmas.

5. The algebra $\mathcal{L}$ defined by

	a	b	c
a	a	a	c
b	a	b	b
c	c	b	c

is SIN in U; this yields $U < W$ in Proposition 71.

6. The algebra $(\mathbb{Z},+)$ is SIN in many varieties. These facts will be proved and applied in Propositions 72, 77, 80 and 81. We could have proved some more of our Propositions with the SINality of $(\mathbb{Z},+)$, but chose alternative arguments having one or another advantage.

7. The groups $\mathbb{Z}_2^n$ and the dihedral groups D_n of order $2n$ are SIN in Gru. This fact is very useful in Proposition 54.

8. The article [28] contains enough information to show that the symmetric groups S_3 and S_4 and the alternating group A_5 are all SIN in Gru. These facts are useful for Proposition 55 below. (The article [28] was prepared before our concept of SIN algebra had properly crystalized; accordingly, it contains more information than is needed for our purposes here.) (Note: $S_3 = D_3$.)

9. The work of showing a finite group to be SIN in Gru is greatly reduced by Proposition 51 in §9. In fact, such are all groups of order ≤ 16, all groups of order 24, and at least 10 of the 13 non-isomorphic groups of order 60.

10. All self-dual lattices are SIN in Lat.

11. The monoid

	1	a	b	c	d
1	1	a	b	c	d
a	a	1	b	c	d
b	b	b	b	c	d
c	c	c	c	c	d
d	d	d	d	d	d

is SIN in the variety of commutative monoids.

12. The monoid

	1	a	b	c	d
1	1	a	b	c	d
a	a	a	a	a	a
b	b	b	b	b	b
c	c	c	c	c	c
d	d	d	d	d	d

is SIN in the variety of monoids.

Non-Examples. Every non-abelian 2-nilpotent group of exponent 3 fails to be SIN in the Burnside variety B_3. (For such a group has an interpretation into an Abelian group of exponent 3, by the formula presented in the introduction.) And obviously, a non-self-dual lattice fails to be SIN in Lat.

Various of our results in the next few sections will reveal more of the structure of L, and a complete picture (such as we know it) was given in pages 7-13. We conclude this section with some observations about the least and greatest elements of L.

Proposition 10. $[V] = 1$ *in* L *iff* $V \models x \approx y$. *And* $[V] = 0$ *in* L *iff* V *has a model of more than one element in which every operation is a projection function.* (I.e., V contains an algebra $\mathfrak{A} = (A; F_s)_{s \in S}$ such that for every s there exists an integer j such that $\mathfrak{A}$ obeys the identity $F_s(x_1, x_2, \ldots) \approx x_j$.) ∎

And thus it is perhaps easiest to think of 1 as $[x \approx y]$ and 0 as $[\emptyset]$ —where by $\emptyset$ we mean the empty set of identities.

Corollary. *If* $[V] > 0$ *then* $[V] \geq [V_0] > 0$ *for some finitely based variety* V_0 *of finite type. For* Σ *an arbitrary finite set of identities, it is decidable whether* $[\Sigma] = 0$, *but not decidable whether* $[\Sigma] = 1$.

Proof. The first sentence follows from a straightforward compactness argument. (For the details, one may consult the proof of Proposition 23 below, of which this is a special case.) To decide whether $[\Sigma] = 0$, one need only test the finitely many functions $s \to j$ (as described above) to see whether one can satisfy the identities Σ by defining $F_s(x_1, x_2, \ldots)$ to be x_j (for all $s \in S$). Finally, we note that in 1967 Perkins proved the impossibility of an algorithm to decide if $\Sigma \vdash x \approx y$ (see [67] or [57]). ∎

2. THE CATEGORICAL POINT OF VIEW

Let C be any category in which finite powers exist, and let A be any object in C. (Our primary examples are only two: A a set, and A a topological space; for the rest of this paper it is necessary to understand this section only in terms of these two examples). For $n = 1,2,3, \ldots$ let A^n denote an nth power of A with projection maps $\pi_i^n : A^n \to A$ $(i = 1, \ldots, n)$. For any map $\lambda : A^m \to A^n$, we let λ_i denote the composite map $\pi_i^n \lambda$ $(i = 1, \ldots, n)$.

Definition. The *variety* V(A) has operations F_μ (of arity n) for each $\mu : A^n \to A$ $(n = 1,2, \ldots)$ and is defined by all laws of the following types:

$$F_{\pi_i^n}(x_1, \ldots, x_n) \simeq x_i$$

$$F_\mu(F_{\lambda_1}(x_1, \ldots, x_m), \ldots, F_{\lambda_n}(x_1, \ldots, x_m))$$

$$\simeq F_{\mu\lambda}(x_1, \ldots, x_m),$$

where $\mu, \lambda, \lambda_1, \ldots, \lambda_n$ are all as above, for any n and any m.

The definition of a topological algebra in V based on a space (A, T) was given in the previous section. We omit the details of the obvious generalization of the next Proposition to category theory.

Proposition 11. *An arbitrary variety* V *has a model in the category* C *based on the object* A (e.g., a topological model with universe the space A) *iff, in the lattice* L, $V \leq V(A)$. ∎

Corollary. $[V(A)]^{[k]} = [V(A^k)]$.

Proof. The inequality $\leq$ is immediate from the Proposition. For the opposite inequality we will interpret $V(A^k)$ in $V(A)^{[k]}$. A typical n-ary operation in $V(A^k)$ is F_μ for $\mu : (A^k)^n \to A^k$. Let $\mu_i = \pi_i^k \circ \mu$, thought of as a kn-ary operation on A; thus operations F_{μ_i} are defined as operations of the variety V(A). Now taking $V(A)^{[k]}$ as presented just before Proposition 6, we may interpret $F_\mu(x_1, \ldots, x_n)$ in $V(A)^{[k]}$ by the term

$$\alpha_\mu(x_1, \ldots, x_n) = d(\alpha_1, \ldots, \alpha_k)$$

where

$$\alpha_i(x_1, \ldots, x_n) = F_{\mu_i}(\lambda_1(x_1), \lambda_2(x_1), \ldots, \lambda_k(x_1), \lambda_1(x_2), \ldots, \lambda_k(x_n)),$$

and where the unary terms $\lambda_i(x)$ $(1 \leq i \leq k)$ are defined in any $V^{[k]}$ as follows:

$$\lambda_i(x) = d(c^{k-i+1}x, c^{k-i+2}x, \ldots, c^{k-i}x).$$

We omit the proof that all the laws of $V(A^k)$ hold under this interpretation. ∎

We remark that the interpretation presented in the last proof is in fact an equivalence of varieties. The Corollary, in fact, gives us a new perspective on the final assertion of Proposition 7. If A is an object in C with $A^k \cong A$, then $[V(A)]^{[k]} = [V(A^k)] = V(A)$. One also easily checks that for U as in Proposition 7 and $A^2 \cong A$, we have $U \leq V(A)$. And U itself is of the form $V(A)$ for an object $A \cong A^2$ in a certain category C; we omit the details--see also the final remarks of this section.

Although we have seen a nice relationship between say, $V(A \times A)$ and $V(A)$, there does not seem to be any similar relation between $V(A \times B)$ and the two varieties $V(A)$ and $V(B)$, other than the obvious relation

$$V(A) \wedge V(B) \leq V(A \times B).$$

For instance, look at the behavior of V_2, V_3 and V_6 described in the next section.

The most important example of $V(A)$ for us in the remainder of this section will arise for C = the category of *sets* and A = some *cardinal* κ. We will in this case denote $V(\kappa)$ by V_κ. Thus in this case, Proposition 11 states that V has a κ-element model iff $V \leq V_\kappa$. For κ a *finite* integer n, the varieties V_κ are well known as the varieties of *Post algebra* or simply the *Post varieties*. From the work of Post, Foster, et al. (see references in Quackenbush [68]), it is known that V_n has finite models in powers n^k $(k = 0,1,2, \ldots)$ and in no other finite powers. Thus the next proposition is immediate.

<u>Proposition 12</u>. $V_n \leq V_m$ *iff* $m = n^k$ *for some* $k \in \omega$. ∎

N.b. The Proposition includes the possibility that $k = 0$, since V_1 is the largest element of the lattice.

<u>Problem</u>. *Is* $V(n^k) \vee V(n^j) = V(n^s)$ *with* $s = \ell.c.m\ (j,k)$? The inequality $\leq$ follows from the Proposition.

Notice that the corresponding assertion for $\wedge$ is easily seen to be false. For example $V_2 \leq V_4 \wedge V_8$, but equality does not hold since $V_4 \wedge V_8$ has no 2-element model. In fact, we will see in §4 that every V_κ is $\bigwedge$-irreducible.

Turning now to look at the varieties V_κ with $\kappa \geq \aleph_0$, we first need a simple lemma of S. Burris and A. Tarski (see [11]). (Of course, if the generalized continuum hypothesis were true, then the lemma would be a trivial consequence of the Löwenheim-Skolem Theorem.)

Lemma. *If* $\lambda \geq \kappa \geq \aleph_0$ *and* V *has a* κ*-element model, then* V *has a* λ*-element model.* ∎

Proposition 13. $V_\kappa < V_\lambda$ *for* $\aleph_0 \leq \kappa \leq \lambda$.

Proof. The Lemma yields $V_\kappa \leq V_\lambda$. And strict inequality holds since there exists a variety with a λ-element model but no κ-element models, e.g., the variety of vector spaces over a field of cardinality λ. ∎

Corollary. L *is a proper class.* ∎

Proposition 14. *If* $V \neq 1$ *and* V *has* $\leq \kappa$ *operations, then* $V \leq V_\kappa$. ∎

Corollary. $1 = \bigvee_{\kappa \in \mathrm{Card}} V_\kappa$. ∎

We close this section with some remarks connected with the category-theoretic material we introduced at the beginning of this section. The rest of the article is independent of these remarks. The construction of $V(A)$ is actually completely general: every variety V is equivalent to $V(A)$ for some object A in some category C. (For instance, one may take C to be the dual of the category V and A to be the V-free algebra on one generator.) Moreover, we may obviously assume that C has distinct objects $A, A^2, A^3, \ldots,$ and no other objects. Such a category C is known as an (abstract) *clone* or an "algebraic theory," according to the Lawvere school. It is not hard to see that our lattice is exactly the same as the lattice which would result from defining $C \leq C'$ for clones to mean that there exists a clone map (i.e., a product-preserving functor) $C \to C'$. In fact, one may think of the defining laws of a variety as being a "presentation" of the corresponding clone, much like a presentation of a group. It is not worthwhile for us to rigidly adopt the abstract clone-theoretic approach in this paper, mainly because we often need to work with these presentations in detail.

3. THE SPECTRUM OF V AND FAILURES OF MODULARITY.

The *spectrum* of a variety V is the set of cardinalities of finite algebras in V:

$$\text{Spec } V = \{n \in \omega^+ : (\exists \mathfrak{A} \in V) \, |\mathfrak{A}| = n\}.$$

It is obvious that Spec V is always a *submonoid* of $(\omega^+; \cdot, 1)$, and in 1967 G. Grätzer proved the converse: every submonoid is Spec V for some V (see [32] or [12, p. 168]; Froemke and Quackenbush later proved [25], that V can be taken to be of type (2)); for finitely generated monoids, one may take V to be the product of V_n, with n ranging over the set of generators). From the previous section, it is clear that

$$\text{Spec } V = \{n \in \omega^+ : V \leq V_n\},$$

for V_n $(n \in \omega^+)$ the Post varieties. From this formula, we clearly have the following:

Lemma. Spec V *is an invariant of* L*; i.e., if* [V] = [W], *then* Spec V = Spec W. ■

The following fact was known to W.D. Neumann [64].

Proposition 15. Spec *defines a lattice homomorphism from* L *onto the dual of the lattice of all submonoids of* $(\omega^+; \cdot, 1)$.

Proof. The fact that

$$\text{Spec } ([V] \vee [W]) = \text{Spec } V \cap \text{Spec } W$$

is immediate from our second expression for Spec V. For $[V] \wedge [W]$, we recall that $[V] \wedge [W] = [V \times W]$; then Proposition 4 tells us that

$$\text{Spec}([V \times W]) = \{mn : m \in \text{Spec } V, \ n \in \text{Spec } W\}.$$

But this last expression is clearly the smallest submonoid containing Spec V and Spec W; i.e.

$$\text{Spec}([V] \wedge [W]) = (\text{Spec } V) \vee (\text{Spec } W).$$

Finally, the homomorphism is onto by Grätzer's theorem mentioned above. ■

Now the *non-modularity of* L will follow immediately from the easily established fact that the lattice of submonoids of $(\omega^{+}; \cdot, 1)$ is not modular. For instance, we have the following submonoids forming a pentagon

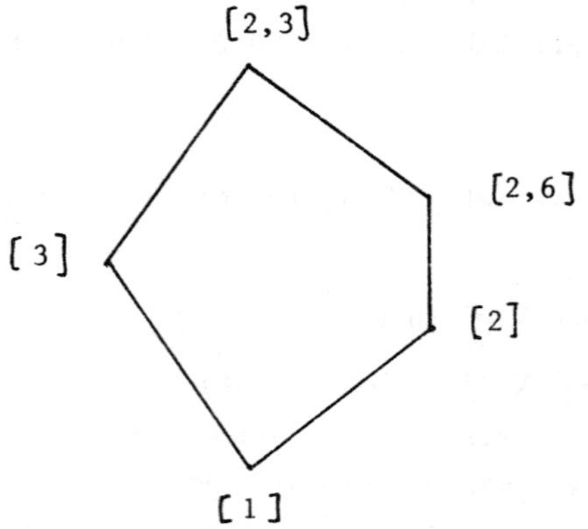

(where, e.g., [2,6] means the smallest submonoid containing 2 and 6, i.e., [2,6] = {1,2,4,6,8,12,16,24,32,36, ...}).

Corollary. L *is not modular.* ∎

Perhaps the simplest pentagon we can directly manufacture in L is

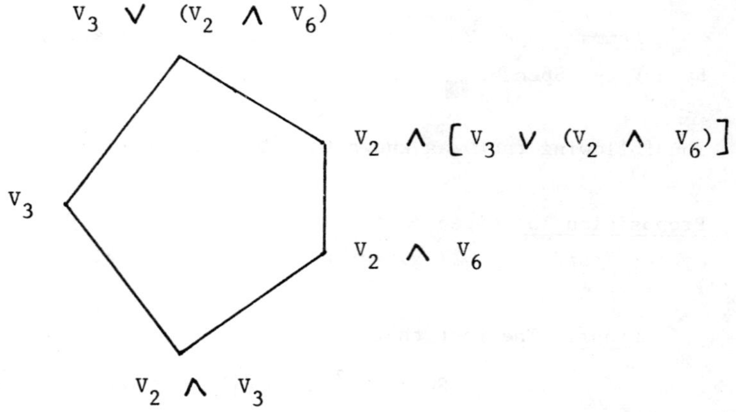

Its spectra are as in the previous diagram (upside down, of course). For a simpler example of failure of the modular law, we have $\{V_2 \wedge V_6\} < V_2$, but

$$[\{V_2 \wedge V_6\} \vee V_3] \wedge V_2 > \{V_2 \wedge V_6\} \vee [V_3 \wedge V_2].$$

(For the RHS has a 6-element model but the LHS does not.)

By considering 32-element models, we easily note the following failure of the distributive law:

$$V_{32} = V_{32} \vee (V_4 \wedge V_8) < (V_{32} \vee V_4) \wedge (V_{32} \vee V_8).$$

(In the next section, we will see that V_{32} is in fact $\bigwedge$-irreducible, even though, as we see here, it fails to be $\bigwedge$-prime. As is well known, every failure of " $\bigwedge$-irrecudible $\to$ $\bigwedge$-prime" involves a failure of the distributive law in this style.) For a simple failure of the other distributive law, by again considering 6-element models, we have

$$V_2 \wedge (V_3 \vee V_6) > (V_2 \wedge V_3) \vee (V_2 \wedge V_6).$$

Problem. Is $V_2 \leq V_3 \vee V_6$?

We remark that the homomorphism of Proposition 15 is not a *complete* homomorphism. In fact, there exist varieties V_i $(i \in \omega)$ such that for each i $\mathrm{Spec}(V_i) = \{1\}$ but $\mathrm{Spec}\,(\bigwedge V_i) = \omega^+$. (See Taylor [79].)

4. $\bigwedge$-PRIME AND $\bigwedge$-IRREDUCIBLE ELEMENTS OF L

An element a of L (or any lattice) is *$\bigwedge$-prime* iff

$$x \wedge y \leq a \rightarrow x \leq a \text{ or } y \leq a$$

holds for all x and y in L. If L satisfies the implication

$$x \wedge y = a \rightarrow x = a \text{ or } y = a,$$

then a is *$\bigwedge$-irreducible.* One easily checks that every $\bigwedge$-prime element is $\bigwedge$-irreducible, and that the converse holds in distributive lattices. In this section we will identify various familiar varieties as $\bigwedge$-prime in L; this information will be applied in the next section to see that various Mal'tsev conditions are prime filters, and hence indecomposable as Mal'tsev conditions.

As we did for Proposition 5, we will state the next proposition for topological spaces only, even though it can be proved more generally (for A an object in a category with equalizers). Proposition 16 certainly applies also to sets (thought of as indiscrete spaces). We will call a space A *primary* iff $A \cong B \times C$ implies $A \cong B^{\kappa}$ or $A \cong C^{\kappa}$ for some cardinal κ.

Proposition 16. *If* A *is a primary space, then* V(A) *is $\bigwedge$-prime in* L.

Proof. Suppose $V \times W \leq V(A)$. Then by Proposition 11, $V \times W$ has a topological algebra with universe A, and by Proposition 5, $A \cong B \times C$ such that V has a topological algebra with universe B and W has a topological algebra with universe C. Without loss of generality, $B^{\kappa} \cong A$, and so V has a topological algebra based on A; hence $V \leq V(A)$. ∎

Other than, of course, the product-indecomposable spaces, primary spaces seem pretty rare. Nonetheless, according to some results of K. Borsuk [8], the Euclidean discs of dimension 2 and 3 are primary, as well as the tori of dimension 2 and 3. Bing [6] proved that the 4-disk is primary. Primary indiscrete spaces are explored in the next Corollary and its proof.

Corollary. *For* κ *a cardinal* > 2, V_{κ} *is $\bigwedge$-prime iff* κ *is infinite or there exists a prime* p *such that* κ *is* p, p^2, p^3, p^4 *or* p^6.

Proof. One easily checks that the cardinals listed here are primary, and so the Proposition yields $\bigwedge$-primeness of these V_{κ}. Moreover, it is easily checked that any finite n not listed is not primal. I.e. $n = st$ such that $n \notin \{s^k : k \in \omega\}$ and $n \notin \{t^k : k \in \omega\}$. Thus $V_s \nleq V_n$, $V_t \nleq V_n$ but $V_s \wedge V_t \leq V_n$ and hence V_n is not $\wedge$-prime. ■

But, we should point out that no V_n $(n < \omega)$ is *infinitely* $\wedge$*-prime.* As was proved in [79], there exist varieties W_i $(i \in \omega)$ with no finite models (i.e., for every i and n, $W_i \nleq V_n$), but with

$$\bigwedge W_i \leq \text{Bounded Distributive Lattices},$$

which has models of every finite power. Thus $\bigwedge W_i \leq V_n$ for every n.

Problem. Is V_ω infinitely $\wedge$-irreducible?

Proposition 17. *Every* V_κ *is* $\wedge$*-irreducible.*

Proof. By the Corollary, we need only prove this for $\kappa = n < \omega$. If $V_n = U \wedge W$, then

$$\{n^k : k \in \omega\} = \text{Spec } U \vee \text{Spec } W.$$

This clearly implies $n \in \text{Spec } U$ or $n \in \text{Spec } W$, i.e., $U \leq V_n$ or $W \leq V_n$. ■

We now present our main general result on $\wedge$-primeness.

Proposition 18. *If* V *is a nested union of varieties* V_i $(i \in \omega)$ *(i.e.,* $V = \bigcup V_i$, $V_0 \subseteq V_1 \subseteq V_2 \subseteq \ldots$), *and each* V_i *is generated by a finite algebra* $\mathfrak{A}_i$ *with a prime number of elements, then* V *is* $\wedge$*-prime in* L.

Remark. Obviously, the hypothesis of a prime number of elements is essential: for instance, $V_2 \wedge V_3$ is generated by a 6-element algebra. But see Corollary 3 to Proposition 22 below, and following remarks, for some methods which sometimes yield $\wedge$-primeness of varieties not generated by prime-power algebras.

Proof. Suppose $U \times W \leq V$. As explained before Proposition 4, we will assume that U is the subvariety of $U \times W$ defined by $xy \approx x$ and W is the subvariety defined by $xy \approx y$. Thus there exists a functor $\Phi : V \to U \times W$ which commutes with the underlying set functor and is defined by V-terms α_t corresponding to the operations F_t of $U \times W$. In particular, let us suppose that the $U \times W$-operation xy is defined by the V-term $\alpha(x,y)$. For each i, by Proposition 4, $\Phi(\mathfrak{A}_i) \cong \mathfrak{B}_i \times \mathfrak{C}_i$ with $\mathfrak{B}_i \in U$ and $\mathfrak{C}_i \in W$. Since $|\Phi(\mathfrak{A}_i)| = |\mathfrak{A}_i|$, which is prime, either $|\mathfrak{B}_i| = 1$ or $|\mathfrak{C}_i| = 1$, and hence

$\Phi(\mathfrak{A}_i) \in U$ or $\Phi(\mathfrak{A}_i) \in W$. In other words, $\mathfrak{A}_i \models \alpha(x,y) \approx x$ or $\mathfrak{A}_i \models \alpha(x,y) \approx y$. Thus w.l.o.g., $V_i \models \alpha(x,y) \approx x$ for infinitely many i. Hence $V \models \alpha(x,y) \approx x$, and so Φ really maps V into U; i.e., $U \leq V$. ∎

We cannot really give a complete list of all the varieties to which the proposition applies, but the Corollary will give some of the better known examples.

Corollary. *The following varieties are* $\wedge$*-prime in* L*:*

Ab_p: Abelian groups of prime exponent p

Semℓ: Semilattices

Lat: Lattices, and more generally, any variety of lattices generated by its finite members

BDL: Bounded distributive lattices

DLC: Distributive lattices with complementation

Mt_0: The majority laws $m(xxy) \approx m(xyx) \approx m(yxx) \approx x$

Mt_1: The Mal'tsev permutability laws
$p(xzz) \approx x \approx (pzzx)$

C: One constant

Binl: $xc \approx cx \approx x$

$\mathcal{U}$: $xy \approx yx$

0: The variety with no operations

Proof. We need to see how, in each case, to find the algebras $\mathfrak{A}_0, \mathfrak{A}_1, \mathfrak{A}_2, \ldots$ of prime order which figure in the hypothesis of the proposition. Now Ab_p, Semℓ, BDL, C and 0 are all generated by a single prime-power algebra, and so we may have $\mathfrak{A}_0 = \mathfrak{A}_1 = \mathfrak{A}_2 = \ldots$, thus settling these cases. There exist finite lattices $L_0, L_1, \ldots$ such that Lat $= HSP\{L_0, L_1, \ldots\}$, for instance, the finite partition lattices (Sachs [70]; see also page 196 of

Grätzer [36]). (The reader may easily devise his own proof using Whitman's Theorem that every lattice is embeddable in a partition lattice.) If we define M_i to be $L_0 \times L_1 \times \ldots L_i$, then each M_i is finite, $\mathrm{Lat} = \mathsf{HSP}\{M_0, M_1, \ldots\}$ and $\mathsf{HSP}\, M_0 \subseteq \mathsf{HSP}\, M_1 \subseteq \mathsf{HSP}\, M_2 \cdots$. Now if we add a finite chain of elements to the top of each M_i, we may ensure that each $|M_i|$ is prime without changing the laws true in M_i. The $\wedge$-primeness of Lat now follows from the Proposition. In the same way, the Corollary follows for any variety of lattices generated by its finite members. (There are $2^{\aleph_0}$ such varieties, see [2], [55] or [36, p. 241]. The argument for pseudocomplemented distributive lattices is much the same.) For the necessary materials consult the expository treatment in Chapter 3 of [35], for the work of K.B. Lee and H. Lakser and G. Grätzer.

Finally, we will describe how to find the algebras $\mathfrak{A}_i$ for the variety U defined by $xy \approx yx$. The only condition required for a multiplication table to define an algebra in this variety is that it be symmetric in the main diagonal. Thus if $(A, \cdot) \in U$ and $B \subseteq A$, then we get a new U-multiplication on B by selecting $b_0 \in B$ and then defining

$$b_1 * b_2 = \begin{cases} b_1 b_2 & \text{if } b_1 b_2 \in B \\ b_0 & \text{otherwise.} \end{cases}$$

In this way we can easily obtain an algebra of prime order which falsifies any finite set of identities which are not consequences of $xy \approx yx$.

The proofs for Mt_0, Mt_1 and Binl are similar. ∎

The previous method will not work for groups; here a similar method yields only $\wedge$-irreducibility. Here G denotes the variety of all groups.

Lemma. *There are only two interpretations of* G *in itself: the identity interpretation, and the one in which* xy *is interpreted by* yx. ∎

Proposition 19. G *is* $\wedge$-*irreducible.*

Proof. We use the easily established fact that $G = \mathsf{HSP}\, G$ for some group G which is directly indecomposable, for instance G = the alternating group on $\aleph_0$ letters. Now suppose that $G = U \times W$ in L. Then there exist universe-preserving functors $\Phi : G \to U \times W$ and $\Psi : U \times W \to G$. By Proposition 4, ΦG is isomorphic to $\mathfrak{A} \times \mathfrak{B}$, with $\mathfrak{A} \in U$, $\mathfrak{B} \in W$. If both $|\mathfrak{A}| > 1$ and $|\mathfrak{B}| > 1$, then $\Psi(\mathfrak{A} \times \mathfrak{B})$ has a proper direct decomposition.

But this is impossible, since $\Psi(\mathfrak{A} \times \mathfrak{B}) \cong \Psi\Phi G$, and by the lemma, $\Psi\Phi G$ is either isomorphic or anti-isomorphic to the indecomposable group G. Thus, w.l.o.g., $|\mathfrak{B}| = 1$ and so $\Phi G \in U$. And so, if xy is interpreted by Φ by the group term $\alpha(x,y)$, it follows that $G \models \alpha(x,y) \approx x$, and hence that $G \models \alpha(x,y) \approx x$. Thus Φ maps G into U, and so $[G] = [U]$. ∎

We omit the proof of the next result; in each case it is proved along the same lines as the previous proposition.

Proposition 20. *The following varieties are $\wedge$-irreducible in* L*: Abelian groups, rings with unit, real vector spaces, rational vector spaces, modular lattices.* ∎

Note that modular lattices do not fall under the rubric of Proposition 18 and its Corollary, since, as R. Freese has proved [22], this variety is not generated by its finite members. (Also see the remarks after Corollary 3 to Proposition 22 below.)

We can get some more precise information about some varieties which are not $\wedge$-prime, by considering a mild generalization of $\wedge$-primeness. We say that an element a of L (or any lattice) has *Helly number* $\leq$ k iff the following holds for all $x_0, \ldots, x_k \in L$:

$$x_0 \wedge \cdots \wedge x_k \leq a \rightarrow x_0 \wedge \cdots \hat{x}_i \cdots \wedge x_k \leq a \quad \text{for some } i,$$

where ^ denotes deletion. The Helly number of a is the least k such that the above property holds. (We thank Joel Berman for coining this term, in honor of Helly's theorem, which states that $\emptyset$ has Helly number n+1 in the lattice of convex subsets of Euclidean n-dimensional space.) Of course, a is $\wedge$-prime iff it has Helly number 1.

Proposition 21. *If* V *is generated by a finite algebra* $\mathfrak{A}$, *and* $|\mathfrak{A}|$ *has exactly* k *prime factors in its prime factorization* (counting repeated primes as often as they occur), *then* V *has Helly number* $\leq$ k.

Proof. An easy generalization of the method of proof of Proposition 18. ∎

Thus for instance every V_n has finite Helly number, although the k provided by Proposition 21 may be unnecessarily large. For instance V_{2i} has Helly number $\leq$ 2 for $i \leq 14$ (although V_{215} has Helly number $\geq$ 3 since

$$V_{24} \wedge V_{25} \wedge V_{26} \leq V_{215},$$

with no reduction possible). Further detailed (or asymptotic) determination of the Helly number of V_n is really pure number theory, and we will not pursue it any further here.

Our main method of estimating Helly numbers in L (and sometimes finding more $\wedge$-prime elements) involves the *free spectrum* of a variety V. We let

$$\omega_n = \omega_n(V) = |F_V(n)|$$

(the cardinality of the V-free algebra on n generators) for $n = 0,1,2, \ldots$. Slightly more information is carried by the cardinal numbers $\alpha_n = \alpha_n(V)$ defined for $n = 0,1, \ldots$ as follows: $\alpha_n(V)$ is the number of n-ary V-operations which depend on all n variables. It is always true that

$$\omega_n = \sum_{i=0}^{n} \binom{n}{i} \alpha_i$$

for every n; this formula holds even if some of the α_i are infinite cardinals. If all ω_i are finite, these equations may be solved for the α_i, yielding

$$\alpha_n = \sum_{i=0}^{n} \binom{n}{i} (-1)^{n-i} \omega_i .$$

We should emphasize that α_n and ω_n are *not invariants* of $[V]$. But this will not matter, for if we can get $\alpha_n(V)$ small (in a sense made precise below), then our Helly number result will hold for $[V]$, even though we will still have $\alpha_n(W)$ infinite for some other $W \in [V]$.

Let S be the variety of "sets" (algebras with no operations), and define S^k to be $S \times \cdots \times S$ (with k factors). (In L, of course, $[S] = [S^k] = 0$, but at present we do care about the distinction between S and S^k.) Finally, we define

$$B(k,n) = \alpha_n(S^k).$$

Lemma.

1. $B(k,n) = \sum_{i=0}^{n} \binom{n}{i} (-1)^{n-i} i^k$

2. $n^k = \sum_{i=0}^{n} \binom{n}{i} B(k,i)$

3. $B(k,n)$ *is the number of functions mapping* k *onto* n. *(In particular,* $B(k,k) = k!$, *and* $B(k,n) = 0$ *for* $n > k$.*)*

4. $B(k,n)$ *is determined by the recurrence*

$$B(k,n) = \sum_{j=n-1}^{k-1} \binom{k}{j} B(j,n-1)$$

together with the initial conditions

$$B(k,1) = 1.$$

5. $B(k,n) = n!S(k,n)$, *where* $S(k,n)$ *is the "Stirling number of the second kind."*

Sketch of proof. It is not hard to see, from the definitions involved, that $\omega_n(S^k) = n^k$. Thus 1 and 2 simply express the equations for α_n's in terms of ω_n's, and vice versa. Now if we define $B'(k,n)$ to be the number of onto functions $k \twoheadrightarrow n$, one easily sees that 2 holds for $B'(k,n)$ in place of $B(k,n)$. From this it easily follows that $B = B'$, thus proving 3. For 4, one can easily check that the numbers $B'(k,n)$ obey the recurrence. It is obvious that the solution to the recurrence with this boundary condition is unique. For the definition of "Stirling number of the second kind," together with enough information to prove 5, see e.g., pages 26-27 of [39]. ∎

Here are a few representative values of our function $B(k,n)$:

k \ n	1	2	3	4	5	6
1	1					
2	1	2				
3	1	6	6			
4	1	14	36	24		
5	1	30	150	240	120	
6	1	62	540	1560	1800	720

Proposition 22. *If* $\alpha_n(V) < B(k+1,n)$, *then the Helly number of* V *in* L *is* $< k$.

Proof. Let us be given an interpretation Φ of $U_0 \wedge \cdots \wedge U_k$ in V. By analogy with Proposition 4, we will assume that $W = U_0 \times \ldots \times U_k$ is our representation of $U_0 \wedge \cdots \wedge U_k$. W has a $(k+1)$-ary operation p such that each U_i is the subvariety of W defined by the law $p(x_0, \ldots, x_k) = x_i$. Moreover, every algebra in W is isomorphic to a product $\mathfrak{A}_0 \times \ldots \times \mathfrak{A}_k$ with each $\mathfrak{A}_i \in U_i$. By way of contradiction, let us suppose that V cannot be interpreted in any $U_0 \times \ldots \hat{U}_i \ldots \times U_k$. Then V must contain, for each i, an algebra such that $\Phi\mathfrak{A}_i = \mathfrak{B}_0 \times \ldots \times \mathfrak{B}_k$ with $|\mathfrak{B}_i| > 1$. Taking $\mathfrak{A} = \mathfrak{A}_0 \times \ldots \times \mathfrak{A}_k$ we have $\phi\mathfrak{A} = \mathfrak{C}_0 \times \ldots \times \mathfrak{C}_k$ with all $|\mathfrak{C}_i| > 1$. Let $\alpha(x_0, \ldots, x_k)$ be the V-polynomial which interprets $p(x_0, \ldots, x_k)$ in W.

We will prove that $\alpha_n(V) \geq B(k+1,n)$ (and thus obtain our contradiction) by substituting variables into $\alpha(x_0, \ldots, x_k)$.

Let λ be an arbitrary onto map $(k+1) \twoheadrightarrow n$. Define $\alpha_\lambda(x_0, \ldots, x_{n-1})$ to be $\alpha(x_{\lambda(0)}, x_{\lambda(1)}, \ldots, x_{\lambda(k)})$. We will prove that if $\lambda \neq \mu$ then α_λ and α_μ are not equivalent in V. For we have $\lambda(i) \neq \mu(i)$ for some i $(0 \leq i \leq k)$. If we had $V \models \alpha_\lambda \approx \alpha_\mu$, then we would have $V \models \alpha(\ldots\ldots,x,\ldots\ldots) \approx \alpha(\ldots\ldots,y,\ldots\ldots)$ (x,y in i^{th} position). But this identity obviously fails on $\mathfrak{A}$, for $\Phi\mathfrak{A}$ cannot satisfy $p(\ldots.,x,\ldots.) \approx p(\ldots.,y,\ldots.)$. This contradiction establishes that all α_λ and α_μ are distinct. A similar argument (omitted) will establish that each α_λ depends on all its variables. Thus we have $\alpha_n(V) \geq$ the number of onto maps $(k+1) \twoheadrightarrow n$, which, by the previous Proposition, is $B(k+1,n)$. ■

<u>Corollary 1</u>. *If* $\omega_n(V) < n^{k+1}$, *then the Helly number of* V *in* L *is* $\leq$ k.

<u>Proof</u>. Immediate from Proposition 22 and clause 2 of the Lemma. ■

<u>Corollary 2</u>. *If* V *is locally finite, or even if* $\alpha_n(V)$ *is finite for one* n, *then* V *has finite Helly number.* ■

<u>Corollary 3</u>. *If* $\alpha_2(V) \leq 1$, *then* V *is* Λ*-prime in* L. ■

It is not hard to find varieties V for which Corollary 3 is applicable, but our previous methods are not. For instance, if A is a finite set and $\mathfrak{A} = (A,T)$ for

$$T(x,y,z) = \begin{cases} z & \text{if } x = y \\ x & \text{if } x \neq y, \end{cases}$$

then it is easily checked that $\mathfrak{A}$ has no non-trivial binary operation. Thus $\alpha_2(V) = 0$ for $V = \mathrm{HSP}\mathfrak{A}$, and hence, by Corollary 3, this V is Λ-prime. Our previous methods (Propositions 16, 18, and 21) do not apply if $|A|$ has more than one prime factor.

The operation T is called the "ternary discriminator" on A, and finite algebras having it among their (derived) operations are called *quasiprimal*. For references on, and a brief introduction to, the theory of such algebras and their generated varieties, we refer the reader to the Quackenbush survey [68]. The main fact we need here is that if $\mathfrak{A} = (A,T)$ and $|A| = n$, then

$$\mathrm{Spec}\ V = \{n^k : k \in \omega\}$$

for $V = \text{HSP}\,\mathfrak{A}$. Now if we let $W_1, W_2, \ldots$ be varieties generated by quasi-primal algebras of distinct prime sizes $p_1, p_2, \ldots$, it should be clear that for each n, $p_1 p_2 \cdots p_n \in \text{Spec}(W_1 \times \ldots \times W_n)$, but $p_1 p_2 \cdots p_n$ is not in the spectrum of any product of $n-1$ of these varieties. Thus $W_1 \wedge \ldots \wedge W_n$ is an irredundant meet in L. On the other hand, one easily checks that $\omega_2(W_i) = 2$, and hence $\omega_2(W_1 \times \ldots \times W_n) = 2^n$, and so, by Corollary 1,

$$W_1 \wedge \ldots \wedge W_n$$

is an irredundant meet with Helly number precisely n.

As a final example of this method, we mention the variety of modular lattices (which already appeared in Proposition 20). One easily sees that this variety has $\alpha_2 = 2 < B(3,2)$, and so has Helly number ≤ 2. Although we already established in Proposition 20 that this variety is $\wedge$-irreducible, we still do not know whether it is $\wedge$-prime.

The only real constraint on the cardinals $\alpha_n(V)$ is the obvious one that $\alpha_1 \geq 1$. G. Grätzer proved in [34] that many cardinal sequences are representable; for example, if $\alpha_0 > 0$, or if $\alpha_0 = 0$ and $\alpha_n > 0$ for $n > 0$ (and $\alpha_1 > 1$), then his proof yields a variety V with $\alpha_n = \alpha_n(V)$. Thus Proposition 22 can be made to apply to many varieties; whether they represent new varieties in L, we do not know. Also see Berman [4]; but his varieties all have a 3-element generic algebra; hence Proposition 18 applies.

A *unary* variety is one whose only operations are unary operations. It is clear that if U is unary, then $\alpha_2(U) = 0$, and hence we have

Corollary 4. *Every unary variety is $\wedge$-prime.* ∎

The natural generalization to ideals of $\wedge$-primeness is the notion of a prime ideal. An ideal I is prime iff

$$x \wedge y \in I \rightarrow x \in I \quad \text{or} \quad y \in I.$$

(And thus a is $\wedge$-prime iff $[0,a]$ is a prime ideal.) Although we will work more with filters than ideals in this article, there is one ideal we can easily see to be prime at this point. Define

$$\mathcal{U} = \{[V] : V \leq U \quad \text{for some unary } U < 1\}$$

It is easy to check that $\mathcal{U}$ is an ideal, since our construction of $\vee$ never introduced any non-unary operations. Note that not every variety in $\mathcal{U}$ is unary, for instance

$$[x+y \approx y+x] \leq [f(x) \approx f(y)],$$

but the smaller one here is not unary. Most of the interesting part of universal algebra seems to take place outside of $\mathcal{U}$.

Corollary 5. $\mathcal{U}$ *is a prime ideal.*

Proof. Immediate from Corollary 4. ■

5. PRIME AND INDECOMPOSABLE FILTERS OF L; AND MAL'TSEV CONDITIONS

It is immediate from Propositions 13 and 14 that 1 is completely $\bigvee$-irreducible in L. (N.b. If L were a set, this would, as usual, imply the existence of a greatest $V < 1$, but L is a proper class, and no such V exists, as can be seen from Proposition 14.) But generally speaking, $\bigvee$-prime elements and $\bigvee$-irreducible elements (defined dually to $\bigwedge$-prime and $\bigwedge$-irreducible elements) are hard to find in L, for reasons made clear in §6 below (although we obtain a few examples at the very end of this section). (For join-irreducibility in another lattice of varieties, see e.g., [20].) Thus it seems reasonable for us to proceed directly to the corresponding notions for filters.

A filter F of L (or any lattice) is called *prime* iff

$$x \vee y \in F \rightarrow x \in F \quad \text{or} \quad y \in F.$$

And F is called *indecomposable* iff F is not the intersection of two larger filters. (And thus, of course, an element $a \in L$ is $\bigvee$-prime [$\bigvee$-irreducible] iff the principal filter $[a,1]$ is prime [indecomposable].) One easily checks that every prime filter is indecomposable, but not conversely.

Filters on L are especially important because of their relationship to *Mal'tsev-definable classes* of varieties. Let us call F a *Mal'tsev filter* in L iff it is a filter generated by countably many finitely based varieties of finite type. Equivalently, iff there exist varieties $V_1 \geq V_2 \geq V_3 \geq \ldots$, each of finite type and finitely based, such that

$$F = \{[W] : (\exists n)\ W \geq V_n\}.$$

A class K of varieties is called *Mal'tsev-definable* iff there exists a Mal'tsev filter F with $K = \bigcup F$ (i.e., $W \in K \leftrightarrow [W] \in F$). The awkward phrase "Mal'tsev condition" seems to mean a sequence $(\Sigma_1, \Sigma_2, \ldots)$ such that Σ_i is a finite set of identities defining V_i above.

Mal'tsev-definable classes assumed an unusual importance because they so often corresponded to natural varietal properties: modularity, distributivity, permutability, regularity, etc., of congruences in the variety, properties of the spectrum, etc. (see [72], [76], [9] and [37], for some examples). It seems that our appreciation of these natural examples can be enhanced if we are able to prove that they represent *indecomposable* filters, and in this section we will be able to do this in some cases of interest. Nevertheless, a couple of tantalizing conjectures remain open. Interestingly, in those cases where we prove a Mal'tsev filter indecomposable, it is indecomposable in the class of all filters, and in those cases in §6 where we prove decomposability, the intersection is actually an intersection of two Mal'tsev classes. We also consider a few filters which are not Mal'tsev filters.

It is well known and easy to prove that if I is a prime ideal in any lattice L, then $L - I$ is a prime filter. Thus we easily obtain the next Proposition and Corollary. (For the definition of ω_n, see just after Proposition 21 above.)

Proposition 23. *If* V *is* $\wedge$*-prime, then*

$$\{W : W \nleq V\}$$

is a prime filter, and if V *is locally finite* (i.e., $\omega_n(V) < \omega$ for all n), *then this prime filter is Mal'tsev definable.*

Proof. We need only prove the second assertion. That is, we must show that if $W \nleq V$, then $W \geq W_0 \nleq V$ for some finitely based variety W_0 of finite type. Let us assume that V and W have disjoint sets of operation symbols, with V defined by identities Σ, and W defined by identities Γ in operation symbols F_t of arity $n(t)$. Taking C_n to be a finite set of V-terms in $x_1, \ldots, x_n$ such that each n-ary term is V-equivalent to a term in C_n, we will consider the compact space $C = \prod_{t \in T} C_{n(t)}$ which results from giving each $C_{n(t)}$ the discrete topology. For each identity $(\sigma \approx \tau) \in \Gamma$, we define $K_{\sigma,\tau}$ to be the subset of C consisting of all $\lambda \in C$ such that

$$\Sigma, \{F_t(x_1, \ldots, x_n) \approx \lambda(t)\} \vdash \sigma \approx \tau.$$

Since the existence of such a deduction depends only on $\lambda(t)$ for F_t appearing in σ and τ, it is clear that $K_{\sigma,\tau}$ is a clopen subset of C.

Now, returning to our main goal, clearly if $W \nleq V$, then no λ can be in every $K_{\sigma,\tau}$. By compactness, there is a finite $\Gamma_0 \subseteq \Gamma$ such that

$$\{K_{\sigma,\tau} : (\sigma \approx \tau) \in \Gamma_0\} = \emptyset.$$

Then it is clear that $W_0 \nleq V$, where W_0 has the operations appearing in Γ_0 and is defined by Γ_0; clearly $W_0 \leq W$. ∎

The next two Corollaries are almost immediate from Proposition 23 and the $\wedge$-primeness results of the previous section.

Corollary 1. *The following classes of varieties are prime Malt'sev classes:*

$$\{W : p \nmid \text{Spec } W\}$$

$$\{W : p^2 \nmid \text{Spec } W\}$$

$$\{W : p^3 \nmid \text{Spec } W\}$$

$$\{W : p^4 \nmid \text{Spec } W\}$$

$$\{W : p^6 \nmid \text{Spec } W\}$$

$$\{W : W > 0\}$$

$$\{W : W \not\leq \text{Semilattices}\},$$

and likewise replacing Semilattices *by* Ab_p *(Abelian groups of exponent* p) *or by the variety generated by a finite quasi-primal algebra* (A;T).

Proof. Each class mentioned here has the form $\{W : W \not\leq V\}$ (e.g., $V = V_p$ for the first one), with V generated by a finite algebra and hence locally finite. Hence, the result follows from Proposition 23 and the previous section. ■

Corollary 2. *In addition to the classes in Corollary* 1, *the following classes of varieties are prime filters in* L:

{W : W has no topological algebra with universe A} (A a primal space)

$$\{W : W \text{ is not } \leq \text{ any unary } V\}$$

$$\{W : W \not\leq \text{Lat}\},$$

and likewise replacing Lat *by any unary variety, or by* $Mt_1 : (p(xzz) \approx x \approx p(zzx))$ *or* $\mathcal{U} : (xy \approx yx)$, *etc.* ■

Thus, for instance, as we saw above, I^4 is primal (I = unit interval), and so

$$\{V : V \text{ has no topological algebra with universe } I^4\}$$

is a prime filter. The same, of course, also holds for any space which is product-indecomposable, such as I itself. It was proved in [76, p. 525] that the prime filter

$$\{V : V \text{ has no topological algebra with universe } I\}$$

is *not* Mal'tsev-definable.

From Corollary 1 we note the heretofore unrecognized Mal'tsev condition

$$\{W : W > \text{Semilattices}\}$$

(*etc.*). This is, of course, the intersection of the principal filter defined by Semilattices with an example from Corollary 1, and so it is not indecomposable or prime. One may easily see that in fact $\wedge$-irreducibility suffices for obtaining this sort of Mal'tsev class, as we see in the next Proposition. Since its proof is so similar to that of Proposition 23, we will omit it.

Proposition 24. *If* V *is* $\wedge$*-irreducible, then*

$$\{W : W > V\}$$

is a filter on L, *and if* V *is locally finite, then this filter is Mal'tsev-definable.* ∎

Corollary. *For each finite* n,

$$\{W : W > V_n\}$$

is a Mal'tsev class of varieties. And

$$\{W : W > \text{Groups}\}$$

is a filter on L *(likewise for Groups replaced by Abelian groups, rings, etc.).* ∎

We do not know whether any filter of the type just mentioned is principal. In fact, we do not know any example of a covering relation holding in L! (Of course if, e.g., the variety of groups had an L-cover W_0, then that cover would be unique and the above filter would be the principal filter generated by W_0.)

Problem 1. Find some (interesting) covering relations in L.

Problem 2. Is $\{W : W > 0\}$ a principal ideal? Equivalently, is there a variety which covers 0 in L? (Note that by the Corollary to Proposition 10, if there is no cover of 0, then $0 = \bigwedge_{i=1}^{\infty} V_i$ for a countable descending chain $V_1 > V_2 > V_3 > \dots > 0$ of finitely axiomatized varieties of finite type.)

The classes defined by the spectrum in Corollary 1 just above are special cases of a more general kind of (Mal'tsev) filter introduced in [72]. For $J \subseteq \omega^+$, the class

$$K_J = \{V : \text{Spec } V \subseteq J\}$$

is a filter if J is a submonoid of $(\omega^+, \cdot, 1)$, and moreover if K_J is a filter then there exists a submonoid* $J' \subseteq J$ such that $K_{J'} = K_J$.

*Taylor mistakenly asserted in [79] that J itself must be a submonoid.

Thus we will concern ourselves only with K_J for J a submonoid. The filter K_J is a Mal'tsev class iff J is cofinite (i.e., $\omega^+ - J$ is finite). None of the filters K_J (for $J \neq \omega^+$) is principal (Taylor [79]). On page 383 of [72] one finds an explicit Mal'tsev condition for Spec V $\subseteq \omega^+ - \{2\}$.

Now Corollary 1 tells us that K_J is a prime filter for J a submonoid of one of the following five types:

$$\omega^+ - \{p\}$$

$$\omega^+ - \{p,p^2\}$$

$$\omega^+ - \{p,p^3\}$$

$$\omega^+ - \{p,p2,p^4\}$$

$$\omega^+ - \{p, p^2,p^4,p^6\}.$$

(For instance, one easily checks that $p^6 \notin$ Spec V iff Spec V $\subseteq \omega^+ - \{p,p^2, p^3,p^6\}$, and that the latter is a submonoid.) In the next proposition we will identify all the prime filters K_J; in particular, we will see that we have already listed all the prime K_J which are Mal'tsev classes.

Proposition 25. *The filter* K_J *is prime iff* $J = \omega^+ - \{p^e : e \in E\}$, *where* p *is prime and* E *is one of the following subsets of* ω^+:

$$\emptyset, \quad \omega^+, \quad \{1\},$$

$$\{1,2\}, \quad \{1,3\},$$

$$\{1,2,4\}, \quad \{1,2,3,6\}, \quad \text{or}$$

$$\{n : q^k \text{ does not divide } n\}$$

for some prime q *and some integer* $k \geq 1$.

Proof. One easily checks that each J listed is a submonoid of $(\omega^+,\cdot,1)$. We content ourselves with proving only that K_J is prime in the last case, namely $J = \omega^+ - \{p^e : e \in E\}$ for $E = \omega^+ - q^k\omega^+$. Now if $V,W \notin K_J$, then V has a p^n-element model and W has a p^m-element model, with q^k dividing neither n nor m. Thus q^k does not divide the least common multiple s of n and m, and $(p^n)^r = p^s = (p^m)^t$ for some r and t. Therefore, V and W each have a p^s-element model, thus yielding a p^s-element model in $V \vee W$. And so we have $V \vee W \notin K_J$, showing that K_J is a prime filter.

Conversely, let us suppose that J is a submonoid of ω^+ such that K_J is prime. First we prove

(*) *If* $m \notin J$, $n \notin J$, *then* $m^s = n^t \notin J$ *for some* s *and* t.

For given $m,n \notin J$, we have $V_m, V_n \notin K_J$, and hence, by the primeness of K_J, $V_m \vee V_n \notin K_J$. Thus

$$J \not\supseteq \text{Spec}(V_m \vee V_n) = \{m^s : s \in \omega\} \cap \{n^t : t \in \omega\},$$

which obviously implies (*). It follows from (*) (and unique prime factorization of integers) that $\omega^+ - J \subseteq \{n^k : k \in \omega\}$ for some n. Moreover, n must be prime, since obviously the *least* member of $\omega^+ - J$ must be prime (otherwise J would not be a submonoid). Thus we have

$$J = \omega^+ - \{p^e : e \in E\}$$

for some prime p and some $E \subseteq \omega^+$. Next we may observe that

(**) $\omega^+ - E$ *is a subsemigroup of* $(\omega^+,+)$,

and

(***) *if* $m,n \in E$, *then* $[m,n] \in E$,

where $[m,n]$ denotes the least common multiple of m and n. (For (**) follows easily from the fact that J is a submonoid, and (***) follows easily from (*) and (**) as follows: (*) yields $ms = nt \in E$ and then (**) yields $[m,n] \in E$.)

One easily checks that an arbitrary subsemigroup of $(\omega^+,+)$ is either cofinite or is contained in $q\omega^+$ for some prime q. (See [54] for a proof.) Thus we have two cases:

Case 1. $\omega^+ - E$ is cofinite, i.e., E is finite. If N is the largest member of E, then E contains every divisor of N by (**), and by (***) every $m \in E$ is a divisor of N (otherwise we would have $[m,N] > N$ and $[m,N] \in E$). Thus we have $E = \{m : m|N\}$, from which one easily sees that $K_J = \{V : V \not\leq V_{p^N}\}$. Since K_J is a prime filter, $\{V : V \leq V_{p^N}\}$ is a prime ideal, i.e., V_{p^N} is $\wedge$-prime in L. It now follows from the Corollary to Proposition 16 that $N = 1,2,3,4$ or 6, and hence a finite E must be one of those appearing on our list.

Case 2. $(\omega^+ - E) \subseteq qN$ for some prime q. Then either $(\omega^+ - E) = \emptyset$ (in which case $E = \emptyset$, which is on our list) or there is a largest k such that $(\omega^+ - E) \subseteq q^k\omega^+$. Thus we must have $q^k w \in \omega^+ - E$ for some w which is not divisible by q. Thus $w \in E$. If we also had $q^k \in E$, then (***) would yield

$$q^k w = [q^k,w] \in E,$$

a contradiction. Hence, $q^k \in \omega^+ - E$ and so

$$q^k\omega^+ \subseteq \omega^+ - E \subseteq q^k\omega^+,$$

i.e., $E = \omega^+ - q^k\omega^+$, which is the final form of E on our list. ∎

The classes K_J also give some interesting examples of indecomposable filters on L which are not prime. Let us look at one easy example before going on to the general result: K_J *is indecomposable for* $J = \omega^+ - \{2,6\}$. Suppose that $K_J = K_1 \cap K_2$ for filters $K_1, K_2 > K_J$. Then for $i = 1,2$, there exists $W_i \in K_i - K_J$. Clearly $V_3 \in K_J$ and so $W_i \wedge V_3 \in K_i$ $(i = 1,2)$. Now W_i has either a 2-element or a 6-element algebra, and so in either case, each $W_i \wedge V_3$ has a 6-element algebra. Thus $W = (W_1 \wedge V_3) \vee (W_2 \wedge V_3)$ has a 6-element algebra. But $W \geq V_i \wedge V_3 \in K_i$, and so $W \in K_1 \cap K_2 = K_J$, which is a contradiction establishing the indecomposability of K_J.

For any cardinal λ we define an element F of a lattice to be λ-*indecomposable* iff F is not the meet of fewer than λ elements larger than F. We are thinking of F in either the lattice of filters on L or the lattice of submonoids of $(\omega^+,\cdot,1)$. Of course F is indecomposable iff it is ω-indecomposable. The next Proposition refers to λ-indecomposability of K_J for arbitrary λ. From the Proposition, one easily infers that ω_1-indecomposability of K_J implies λ-indecomposability for all $\lambda \geq \omega_1$, and hence the notion of λ-indecomposability of K_J is redundant for $\lambda \geq \omega_2$. Nevertheless, one cannot see this without first proving the Proposition for arbitary λ.

Proposition 26. K_J *is* λ-*indecomposable iff* J *is* λ-*indecomposable in* $\mathrm{Sub}(\omega^+,\cdot,1)$.

Proof. Suppose that J is not λ-indecomposable, i.e., that $J = \bigcap_{i<\kappa} J_i$ for some submonoids $J_i > J$ and some $\kappa < \lambda$. It is easy to see that $K_J = \bigcap_{i<\kappa} K_{J_i}$ and that each $K_{J_i} > K_J$.

Conversely, suppose that K_J is not λ-indecomposable, i.e., that

$$K_J = \bigcap_{i<\kappa} K_i$$

for some $\kappa < \lambda$ and some filters $K_i > K_i$. For each $i < \kappa$ let W_i be a variety in $K_i - K_J$; thus, $\mathrm{Spec}\, W_i \nsubseteq J$. By Grätzer's Theorem cited above, there exists a variety W with Spec W = J. Now replacing each W_i by $W \wedge W_i$, we have $W_i \in K_i$ (since K_i is $\wedge$-closed) and $\mathrm{Spec}\, W_i > J$. To finish, we will prove that

$$J = \bigcap_{i<\kappa} \mathrm{Spec}\, W_i$$

The inclusion $\subseteq$ is evident; for the opposite inclusion, we take $m \in \bigcap \mathrm{Spec}\, W_i$ and prove that $m \in J$. Each W_i has an m-element model, and hence so does $\bigvee W_i$. But $\bigvee W_i \geq$ each W_i, and hence it is in each K_i, and hence in

K_J. Thus, K_J contains a variety (namely W_1) with an m-element model; thus $m \in J$. ∎

Corollary. *The following conditions are equivalent:*

(1) K_J *is* λ*-indecomposable for all* λ;

(2) K_J *is* ω_1*-indecomposable;*

(3) K_J *is indecomposable and* $\omega^+ - J$ *is finite.*

Proof. By the Proposition, it suffices to prove the equivalence of these conditions with K_J replaced by J. Certainly indecomposability of a cofinite set implies λ-indecomposability for all λ; hence (3) → (1). And (1) → (2) a fortiori, and so it remains to prove (2) → (3). All that is needed is to show that if $\omega^+ - J$ is infinite, then J is not ω_1-indecomposable. It is easy to see that $J_n = J \cap \{m : m \geq n\}$ is a submonoid and that $J = \bigcup J_n$. Clearly, if $\omega^+ - J$ is infinite, then each $J_n > J$ and so J is not ω_1-indecomposable. ∎

It does not seem feasible to us to make a complete catalog of all indecomposable filters K_J (as we did for the prime case in Proposition 25). Instead we content ourselves with a few examples:

J	Properties of K_J
$\omega^+ - \{2,4\}$	Prime and ω_1-indecomposable.
$\omega^+ - \{2,4,8,16,\ldots\}$	Prime and indecomposable, but not ω_1-indecomposable.
$\omega^+ - \{2,6\}$	ω_1-indecomposable, but not prime.
$\{1,3,5,7,9,\ldots\}$	Indecomposable, but neither prime nor ω_1-indecomposable.
$\omega^+ - \{2,3\}$	Not indecomposable.

Now we examine another kind of Mal'tsev class which can be shown prime by examining its models. Here we will examine not the cardinalities of the models, but rather their congruence lattices. Here we have in mind the class of *congruence-regular* varieties. An algebra is called congruence-regular iff it does not have two distinct congruences with a block in common. A variety V is congruence-regular iff every algebra in V is congruence-regular. Or equivalently (Thurston [80]), V is congruence-regular iff no algebra in V has a non-trivial congruence with a singleton block. It was shown independently by

Csákány, Grätzer and Wille that the class of all congruence-regular varieties is Mal'tsev-definable. For references, together with another (indirect) argument, see pages 377-378 of Taylor [72].

Proposition 27. *The Mal'tsev filter of regular varieties is prime.*

Proof. We must simply prove that if V is not congruence-regular and W is not congruence-regular, then V $\vee$ W is not congruence-regular. By Thurston's result cited above, we may assume that there exists an algebra $\mathfrak{A} \in V$ and a congruence θ on $\mathfrak{A}$ which has at least one block of one element and at least one block of more than one element. And there exists a similar congruence ψ on an algebra $\mathfrak{B} \in W$. Let κ be the larger of $|\mathfrak{A}|$ and $|\mathfrak{B}|$. Define $\mathfrak{C} \subseteq \mathfrak{A}^{\kappa}$ as follows:

$$C = \{\alpha \in \mathfrak{A}^{\kappa}\colon (\forall i,j < \kappa)\alpha_i \theta \alpha_j\}.$$

It is clear that $\mathfrak{C}$ is a subalgebra of $\mathfrak{A}^{\kappa}$ of power 2^{κ}. Now if ϕ is the congruence on $\mathfrak{C}$ given by $\alpha\phi\beta \leftrightarrow \alpha_0\theta\beta_0$, it is clear that the blocks of ϕ are of power 1 and 2^{κ}, precisely. Now, multiplying by an algebra in V of power 2^{κ}, we may be sure that ϕ has 2^{κ} singleton blocks, 2^{κ} blocks of power 2^{κ}, and nothing else.

Applying a similar construction within W, we finally arrive at the following situation: algebras $\mathfrak{A} \in V$, $\mathfrak{B} \in W$ and congruences $\theta \in \mathrm{Con}\,\mathfrak{A}$, $\psi \in \mathrm{Con}\,\mathfrak{B}$, and having 2^{κ} singleton blocks, 2^{κ} blocks of power 2^{κ} and no other blocks. Now there clearly exists a one-one correspondence $\lambda : \mathfrak{A} \to \mathfrak{B}$ which maps the equivalence relation θ to the equivalence relation ψ. Thus if we identify the universe of $\mathfrak{A}$ with the universe of $\mathfrak{B}$ according to the bijection λ, we obviously obtain an algebra in V $\vee$ W which does not have regular congruences. ∎

There are some other Mal'tsev conditions which may be seen to be prime with reasoning like that of the proof of Proposition 27. We will mention a few, but we will not feel it necessary to include every detail of their proofs. Our first example is that of semi-degeneracy, introduced by Lausch and Nöbauer in [51] (see especially pages 13-14, 42, and 55-56). A variety V is *semi-degenerate* iff no algebra in V (other than the one-element algebra) has a one-element subalgebra. The semi-degenerate varieties obviously form a filter in L; it is a Mal'tsev filter according to a result of B. Csákány [15], who gave explicit Mal'tsev conditions. Interestingly, V is semi-degenerate iff every algebra in V has a compact unit element in its congruence lattice. (J. Kollár [48]; see also R. W. Quackenbush [69]). On page 59 of [75] (see also [15]), it was proved that there exists a semi-degenerate variety which does not lie above C = the variety with one constant. In other words, the filter of semi-degenerate varieties is not contained in the principal filter generated by C. This fact is now completely evident from the next Proposition and Corollary 3 to Proposition 36 below.

Proposition 28. *The Mal'tsev filter of semi-degenerate varieties is prime.* ∎

A variety V is *weakly congruence-uniform* iff for every cardinal m there exists a cardinal n such that whenever B_1 and B_2 are blocks of a congruence on an algebra in V, if $|B_1| \leq m$, then $|B_2| \leq n$. For more information on this notion, see [73]. It defines a Mal'tsev filter strictly smaller than the filter of congruence-regular varieties.

Proposition 29. *The Mal'tsev filter of weakly congruence-uniform varieties is prime.*

Proof. We will temporarily invoke the generalized continuum hypothesis (GCH). Since specific generators for this Mal'tsev condition can be seen in [73], it is clear that Proposition 29 is equivalent (without GCH) to a very simple proposition about some deductions in equational logic, and hence, is *absolute*. (For the theory of absoluteness, adequate to establish the above assertion, see e.g., pp. 21-24 of Jech [42].) It then follows that GCH can in principle be removed from our proof. (For some other examples of this style of proof by eliminating the GCH, see Keisler [46] or Galvin [26; §6]; also see Kreisel [49].)

As in the previous proof, we will prove that if neither V nor W is weakly congruence-uniform, then neither is $V \vee W$. Taking n to be an infinite cardinal $\geq$ the number of operations in $V \vee W$, we know that there exists $\mathfrak{A} \in V$ and a congruence θ on $\mathfrak{A}$ which has blocks of power n and $\aleph_\omega(n)$ (the ω-th cardinal larger than n). Presently, we will show how to modify $\mathfrak{A}$ and θ so that θ will have exactly $\aleph_\omega(n)$ blocks of each power n, $\aleph_1(n) = n^+$, $\aleph_2(n) = n^{++}$, ..., $\aleph_\omega(n)$, and no block of any other size. Such a construction can also be performed within the variety W, and so we obtain (in the manner of the previous proofs) an algebra in $V \vee W$ having a congruence with blocks of powers n and $\aleph_\omega(n)$. Recall that the numbers $\beth_i(n)$ and $\beth_\omega(n)$ are defined recursively as follows:

$$\beth_0(n) = n$$

$$\beth_{i+1}(n) = 2^{\beth_i(n)}$$

$$\beth_\omega(n) = \sum_{i<\omega} \beth_i(n)$$

Obviously GCH implies that $\beth_\omega(n) = \aleph_\omega(n)$, and so, assuming GCH, we have an algebra $\mathfrak{A} \in V \vee W$ and a congruence ψ on $\mathfrak{A}$ which has blocks of power n and $\beth_\omega(n)$. It now follows from Theorem 1.6 of [73] that $V \vee W$ is not weakly congruence-uniform.

And so it remains for us to present the promised modification of $\mathfrak{A}$ and θ to yield the correct block sizes. Taking a subalgebra will always reduce all blocks to $\aleph_\omega(n)$ without increasing any block; and taking a product with an algebra of power n will increase each block B to a block of power $|B| + n$ (using the congruence $\theta \times 1$). Moreover, taking a product with an algebra of power $\aleph_\omega(n)$ (and using the congruence $\theta \times 0$) will replicate each block $\aleph_\omega(n)$ times. And thus all we need to do now is to modify $\mathfrak{A}$ and θ so that θ will have at least one block of each power $\aleph_i(n)$ $(i = 0,1,2,\ldots,\omega)$.

We are given a congruence θ on $\mathfrak{A} \in V$ with blocks C and B such that $|C| = n$ and $|B| = \aleph_\omega(n)$. Clearly, we can find an increasing sequence of subalgebras

$$\mathfrak{A}_0 \subseteq \mathfrak{A}_1 \subseteq \mathfrak{A}_2 \subseteq \ldots$$

such that $|C \cap A_i| = n$ and $|B \cap A_i| = \aleph_i(n)$ $(i \in \omega)$. Let $\mathfrak{B}$ be the subalgebra of the product $\prod \mathfrak{A}_i$ consisting of all sequences which are constant beyond a certain point, and define a congruence ϕ on this as follows:

$$(a,b) \in \phi \leftrightarrow \forall i(a_i,b_i) \in \theta.$$

Now choose $c \in C$ and $b \in B$. It is clear, for each i, that the ϕ-block of

$$(\underbrace{c,\ldots,c}_{\leftarrow\ i\ \rightarrow},b,c,c,\ldots)$$

contains exactly $\aleph_i(n)$ elements. Finally, the congruence of $\theta \times \phi$ on $\mathfrak{A} \times \mathfrak{B}$ has blocks of every power $\aleph_i(n)$ $(0 \leq i \leq \omega)$. ∎

A prime Mal'tsev filter which is yet again smaller than the one just studied is Geiger's class of *coherent* varieties (see [73]). Note that in fact the class of coherent varieties is also contained in the class of congruence-permutable varieties:

$$\mathcal{C} \subseteq \mathcal{WU} \cap \mathcal{CP}$$

where $\mathcal{C}$, $\mathcal{WU}$ and $\mathcal{CP}$ are the Mal'tsev filters corresponding to coherence, weak uniformity and permutability, respectively. It was conjectured by Chajda that equality holds here, but this conjecture was disproved by Taylor around 1980. See Chajda [13] for Taylor's rather complicated counterexample. In fact it is much easier to simply apply primality, which is our next result.

A variety V is called *coherent* iff the following condition holds: if $\mathfrak{A} \subseteq \mathfrak{B} \in V$, ϕ is a congruence on $\mathfrak{B}$, and $\mathfrak{A}$ contains one block of ϕ, then $\mathfrak{A}$ is a union of blocks of ϕ.

Proposition 30. *The Mal'tsev filter of coherent varieties is prime.* ∎

Instead of trying to find every last example of a filter which can be proven prime in the manner of Proposition 27, we will conclude this line of reasoning with some final results which in fact give us our only examples so far of prime filters (other than {1}) which are *principal*. That is, we can now give a few examples of $\vee$-prime elements of the lattice L. Let us recall that H.-P. Gumm proved in [37] that an arbitrary variety V has a binary term + obeying $x+y \approx y+x$ iff every involutory automorphism of every algebra in V has a fixed point. (An automorphism $\phi : \mathfrak{A} \to \mathfrak{A}$ is called involutory iff $\phi^2 = \phi$.) But the filter F of all such varieties is clearly prime, by reasoning that by now is familiar. For if $V \notin F$, then V has a model $\mathfrak{A}$ which can be partitioned into two-element sets such that one obtains an automorphism by switching the elements of each pair. Likewise, $W \notin F$ has an algebra with such a partition. We easily modify $\mathfrak{A}$ and $\mathfrak{B}$ to be of the same cardinality, and hence we may patch them to form an algebra in $W \vee V$, which has an involutory automorphism with no fixed point. Since F is the principal filter generated by $[x+y \approx y+x]$, we have

Proposition 31. $[x+y \approx y+x]$ *is* $\vee$*-prime in* L. ∎

And in exactly the same manner, using Gumm [37], we may establish

Proposition 32. *For any prime* p, $[F(x_1, \ldots, x_p) \approx F(x_2, \ldots, x_p, x_1)]$ *is* $\vee$*-prime in* L. ∎

If we temporarily let W_p denote the variety just defined (for p not necessarily prime), then it is clear that

$$W_{mn} = W_m \vee W_n$$

for m and n relatively prime, and so Proposition 32 will not extend to all integers. We are not sure of the situation regarding W_{p^k} (p prime).

Finally, we note that an arbitrary variety V has a constant operation iff any two subalgebras of any algebra in V have a point in common (see [71], [72]). From this fact, it is not hard to see the following result, where, as above, C denotes the variety with one constant operation.

Proposition 33. C *is* $\vee$*-prime in* L. ∎

Our only method for proving $\vee$-primeness of an element and primeness of a filter seems to be the "model-theoretic" method exemplified throughout this section. It doubtless would be fruitful to also have available a more syntactic method for proving $\vee$-primeness, etc., that is, a method that somehow would proceed by examining the *formulas* defining, say, an interpretation of U in V $\vee$ W.

Problem. *Obtain a syntactic method for proving* $\vee$*-primeness or* $\vee$*-irreducibility, etc.*

We would probably expect such a method to be strong enough to prove, e.g., Propositions 32 and 33. In most of the other results of this section, the associated Mal'tsev conditions seem too complicated to allow a workable syntactic method. Nevertheless, what we would like for such a method is for it to be sufficiently strong to answer "yes" to the two outstanding problems in this area which we now state. (We conjecture that both of these problems have affirmative solutions.)

Problem. *Do congruence-modular varieties form a prime Mal'tsev filter?*

Problem. *Do congruence-permutable varieties form a prime Mal'tsev filter?*

Of course, this last filter is the principal filter generated by $[p(xzz) \approx x,\ p(xxz) \approx z]$, according to the result of Mal'tsev which gave this topic its name. The particularly simple form of this last pair of identities lends some hopefulness to the idea of finding a syntactic solution. The same seems true for modularity, given the relatively simple Mal'tsev conditions by Day [18] and Gumm [38].

6. SOME FILTERS WHICH ARE NOT INDECOMPOSABLE OR NOT PRIME.

Propositions 25 and 26 above (and the remarks following Proposition 32) gave some examples of filters which fail to be prime or even fail to be indecomposable. Here we will examine some other examples, notably the filter of congruence-distributive varieties and the principal filter generated by lattice theory. But first, for comparison with Proposition 29, we will look at the class of *congruence-uniform* varieties, i.e., varieties V such that every congruence on every $\mathfrak{A} \in V$ is a partition of $\mathfrak{A}$ into equinumerous blocks.

<u>Proposition 34</u>. *The filter of congruence-uniform varieties is not prime.*

<u>Proof</u>. This is almost immediate from the discussion of Σ on pages 353-355 of [73]. The Σ described there is a union of two sets of identities in disjoint similarity types. The first set of identities forces congruence-blocks to be infinite, and the second forces uniformity to hold when the blocks are infinite. For details, we refer the reader to [73]. ∎

R. Freese informed us in 1980 that the filter of congruence-distributive varieties is not indecomposable, for it is clearly the intersection of the congruence-modular filter with the filter of varieties V such that no algebra V has congruence lattice with a sublattice isomorphic to the diamond 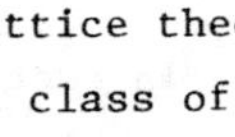 .

The latter filter is strictly larger than the congruence-distributive filter, since it contains the variety of semilattices, by a result of D. Papert--see [23]. Whether it is a Mal'tsev filter, we do not know. Nevertheless, we have the following result.

<u>Proposition 35</u>. *The filter of congruence-distributive varieties is the proper intersection of two Mal'tsev filters.*

<u>Proof</u>. We will work exclusively with finitely based varieties and Mal'tsev conditions in this proof. In 1967, B. Jónsson gave [43] some explicitly finitely axiomatized varieties D_N $(N = 1,2,\ldots)$ such that $D_1 \geq D_2 \geq \ldots$ and such that any V is congruence-distributive iff $V \geq D_N$ for some N. See also Jónsson [44]. Our proof will be complete as soon as we have constructed some finitely axiomatized varieties P_N, Q_N and we have shown them to have the

following properties

$$P_1 \geq P_2 \geq P_3 \geq \cdots$$

$$Q_1 \geq Q_2 \geq Q_3 \geq \cdots$$

$$P_N \vee Q_N = D_N \qquad (N = 1,2,\ldots)$$

P_1 not congruence-distributive

Q_1 not congruence-distributive.

In fact, it will be simpler notationally for us to do this *only for odd* N, even though it can be done for all N. This shortcut is sufficient, since clearly any infinite subsequence of $\{D_N\}$ defines the same Mal'tsev condition. And for odd N, Jónsson's D_N may be taken to have ternary operations m_i $(1 \leq i \leq N)$ and to be axiomatized as follows:

$$m_i xyx \approx x \qquad (\text{all } i)$$

$$m_1 xxz \approx x$$

$$m_i xzz \approx m_{i+1} xzz \qquad (i \text{ odd})$$

$$m_i xxz \approx m_{i+1} xxz \qquad (i \text{ even})$$

$$m_N xzz \approx z$$

Now we take our P_N to have 6-ary operations p_i $(1 \leq i \leq N)$ and to be axiomatized as follows:

$$p_i xyxwxw \approx x \qquad (\text{all } i)$$

$$p_1 xxzxww \approx x$$

$$p_i xzzuuw \approx p_{i+1} xzzvvw \qquad (i \text{ odd})$$

$$p_i xxzuvv \approx p_{i+1} xxzuww \qquad (i \text{ even})$$

$$p_N xzzuuz \approx z.$$

Clearly, P_{N+2} is interpretable in P_N by defining p_i' to be p_i $(1 \leq i \leq N)$ and also p_{N+1}' and p_{N+2}' to be p_N. And P_N is definable in D_N by defining $p_i xyzuvw$ to be $m_i xyz$. And to see that P_1 does not define a congruence-distributive variety, we need only note that $P_1 \leq A_2$ (the variety of (Abelian) groups of exponent 2), as follows: define $p_1 xyzuvw$ to be $u + v + w$.

Our Q_N has ternary operations α_i, β_i and γ_i, $(1 \leq i \leq N)$ and is defined by the following identities:

$$\alpha_i xyx \approx \gamma_i xyx \qquad \beta_i xyx \approx x \qquad \text{(all } i)$$

$$\alpha_1 xxz \approx x$$

$$\left.\begin{aligned} \alpha_i xzz &\approx \beta_i xzz \\ \beta_i xxz &\approx \gamma_i xxz \end{aligned}\right\} \text{ all } i$$

$$\gamma_i xzz \approx \gamma_{i+1} xzz \qquad (i \text{ odd})$$

$$\alpha_i xxz \approx \alpha_{i+1} xxz \qquad (i \text{ even})$$

$$\gamma_N xzz \approx z.$$

It is easy to check that $Q_{N+2} \leq Q_N$ by reasoning similar to that given above for $P_{N+2} \leq P_N$. And Q_N is interpretable in D_N by defining $\alpha_i = \beta_i = \gamma_i = m_i$ $(1 \leq i \leq N)$. And to see that Q_1 does not define a congruence-distributive variety, we need only note that $Q_1 \leq Sem\ell$ (the variety of semilattices), as follows:

$$\text{define } \alpha_1 xyz \text{ as } x \vee y$$

$$\text{define } \beta_1 xyz \text{ as } x \vee z$$

$$\text{define } \gamma_1 xyz \text{ as } y \vee z.$$

To complete the proof we need only show that $D_N \leq P_N \vee Q_N$ for each N. We define

$$m_i xyz = p_i(x,y,z,\alpha_i xyz,\beta_i xyz,\gamma_i xyz)$$

for $i = 1,\dots,N$, and then check the Jónsson axioms for D_N, as follows:

$$m_i xyz \approx p_i(x,y,x,\alpha_i xyx,\beta_i xyx,\gamma_i xyx)$$

$$\approx p_i xyxwxw \approx x.$$

$$m_1 xxz \approx p_1(x,x,z,\alpha_1 xxz,\beta_1 xxz,\gamma_1 xxz)$$

$$\approx p_1 xxzxww \approx x.$$

For i odd,

$$m_i xzz \approx p_i(x,z,z,\alpha_i xzz,\beta_i xzz,\gamma_i xzz)$$

$$\approx p_i(x,z,z,u,u,\gamma_i xzz)$$

$$\approx p_{i+1}(x,z,z,v,v,\gamma_{i+1}xzz)$$

$$\approx p_{i+1}(x,z,z,\alpha_{i+1}xzz,\beta_{i+1}xzz,\gamma_{i+1}xzz)$$

$$\approx m_{i+1}xzz$$

(with u standing for $\alpha_i xzz \approx \beta_i xzz$ and v standing for $\alpha_{i+1}xzz \approx \beta_{i+1}xzz$). The corresponding identity for even i has a similar proof. And finally,

$$m_N xxz \approx p_N(x,z,z,\alpha_N xzz,\beta_N xzz,\gamma_N xzz)$$

$$\approx p_N xzzuuz \approx z. \blacksquare$$

Thus, in particular, we see that the "majority identities" Mt_0:

$$m(xxy) \approx m(xyx) \approx m(yxx) \approx x$$

form a variety which is not $\vee$-irreducible. (It was shown to be $\wedge$-prime in §4).

Given that the filter D of congruence-distributive varieties is $P \cap Q$ with P and Q the filters defined in the last proof, we still do not know whether P is modular. (Although if congruence-modularity defines a prime filter, as we suspect it may, then P must be modular.) In any case, if M denotes the congruence-modular filter, then $D = (P \cap M) \cap (Q \cap M)$, with $P \cap M > D$. Thus, either $Q \cap M = D$ or D is the proper intersection of two subfilters of M.

To substantiate our earlier claim that there are "few" $\vee$-prime and $\vee$-irreducible elements of L, we will now prove that *no $\vee$-irreducible variety has a constant operation* (except of course for $[V] = 1$ and $[V] = [C]$). Here C will denote the variety with one constant operation and no other operations.

For this purpose we introduce the notion of the *parametrization* V_0 of an arbitrary variety V. If V has operations F_t $(t \in T)$ (each of arity n_t) and is defined by the identities $\sigma_i \approx \tau_i$ $(i \in I)$, then we take V_0 to have operations F_t^+ $(t \in T)$ (each of arity $n_t + 1$) and to be defined by the identities $\sigma_i^+ \approx \tau_i^+$ $(i \in I)$. Here the V_0-term α^+ is defined for each V-term α by introducing a new variable z and recursively defining

$$x_i^+ = x_i$$

$$(F_t(\alpha_1,\ldots,\alpha_n))^+ = F_t^+(z,\alpha_1^+,\ldots,\alpha_n^+).$$

Model-theoretically the description of V_0 is clear: if $(A,F_t^+) \in V_0$, then for each choice of "parameter" $a \in A$, we obtain an algebra $(A,F_t^a) \in V$ by defining $F_t^a(x_1,\dots,x_n)$ to be $F_t^+(a,x_1,\dots,x_n)$. (And thus for $a \neq b$, the operations F_t^a and F_t^b are not related in any way.) Moreover, all algebras in V_0 have this form.

Proposition 36.

1. $V_0 \le V$ *and* $V_0 \le V_{00}$.

2. *If* $V \le W$, *then* $V_0 \le W_0$, *and so* $[V]_0 = [V_0]$ *is well defined in* L.

3. $(V \wedge W)_0 \le V_0 \wedge W_0$ *in* L.

4. $(V \vee W)_0 = V_0 \vee W_0$ *in* L.

5. $V \le V_0 \vee C$.

6. *If* $[V] \neq 1$, *then* $V_0 \not\geq [x+y \approx y+x]$.

7. *If* $[x+y \approx y+x] \le [V] < 1$, *then* $[V_0] < [V]$.

Remark. Thus, the map $V \longrightarrow V_0$ is a retraction of the semilattice $(L, \vee)$. And thus 6 (together with Proposition 31) yields another proof of the $\vee$-indecomposability of 1 in L.

Proof of Proposition 36. We omit the proofs of the rather obvious parts 2, 3 and 4, and we remark that 7 is clearly a corollary of 1 and 6. Thus it remains to prove 1, 5 and 6. It is obvious that $V_0 \le V$; for the other half of 1, we observe that the operations of V_{00} have the form $F_t^{++}(z',z,x_1,\dots,x_n)$. Now we may simply interpret $F_t^+(z,x_1,\dots,x_n)$ as $F_t^{++}(z,z,x_1,\dots,x_n)$, yielding $V_0 \le V_{00}$.

To prove 5, we note that if c is a constant, we may interpret each $F_t(x_1,\dots,x_n)$ as $F_t^+(c,x_1,\dots,x_n)$. For 6, suppose that $V < 1$, and hence that V has an infinite model $(A;F_t)_{t \in T}$. Thus, there exists a bijection $\phi : A \to A$ with no fixed points such that $\phi^2 \neq 1$. (One constructs ϕ by partitioning A into two-element blocks and then declaring each block to be a ϕ-orbit.) We may write $A = A_1 \cup A_2$, with $A_1 \cap A_2 = \emptyset$ and such that $\phi[A_1] = A_2$. Now for each $t \in T$, we define our $(n_t + 1)$-ary operation F_t^+ on A as follows:

$$F_t^+(b,a_1,\dots,a_n) = \begin{cases} F_t(a_1,\dots,a_n) & \text{if } b \in A_1 \\ \phi^{-1}F_t(\phi a_1,\dots,\phi a_n) & \text{if } b \in A_2. \end{cases}$$

It should be clear that $(A;F_t^+)_{t \in T} \in V_0$. We claim, moreover, that ϕ is an automorphism of this algebra, which we see as follows:
if $b \in A_1$, then

$$\begin{aligned}\phi F_t^+(b,a_1,\ldots,a_n) &= \phi F_t(a_1,\ldots,a_n)\\ &= \phi^{-1}F_t(\phi^2 a_1,\ldots,\phi^2 a_n)\\ &= F_t^+(\phi b,\phi a_1,\ldots,\phi a_n).\end{aligned}$$

And if $b \in A_2$, then

$$\begin{aligned}\phi F_t^+(b,a_1,\ldots,a_n) &= \phi^2 F_t(\phi a_1,\ldots,\phi a_n)\\ &= F_t(\phi a_1,\ldots,\phi a_n)\\ &= F_t^+(\phi b,\phi a_1,\ldots,\phi a_n).\end{aligned}$$

Thus, V_0 has an algebra which has an involutory automorphism with no fixed points. According to the result of H.-P. Gumm, described on page 57, $V_0 \geq [x+y \approx y+x]$. ∎

In the following Corollaries, F will denote a non-trivial filter on L, i.e., a filter $F > \{1\}$. We call F *0-closed* iff $V_0 \in F$ whenever $V \in F$. Here F_2 will denote the Gumm filter of varieties $V \geq [x+y \approx y+x]$, and $\mathbf{C}$ will denote the filter of varieties with at least one constant operation.

Corollary 1. *If* F *is prime, then* $F \supseteq \mathbf{C}$ *or* F *is* 0-*closed.*

Proof. If $F \not\supseteq \mathbf{C}$, then $C \notin F$. Now for any $V \in F$, Part 5 of the Proposition tells us $V \leq V_0 \vee C$. Now we must have $V_0 \in F$, for otherwise F would fail to be prime. ∎

Of course the prime filter $F = L$ satisfies both of the conclusions of Corollary 1, and the prime filters $\mathbf{C}$ and F_2 satisfy only the first conclusion. But most of the more familiar filters fail to satisfy the first conclusion. Thus, for instance, we see that regularity, coherence, exclusion of p from the spectrum, etc., are all 0-closed filters. In fact, in all these cases, the 0-closedness can easily be verified directly. The reader may be interested in checking that D (the congruence-distributive filter) is also 0-closed, even though it is a proper intersection as we saw above. Thus, the Corollary will not be useful to us for proving that familiar filters or Mal'tsev conditions fail to be prime. But, as we will see in further Corollaries, the real utility of this method is in applications to *principal* filters, i.e., in proving that some familiar varieties fail to be $\vee$-prime.

Corollary 2. *If* F *is prime, then* $F \supseteq \mathbf{C}$ *or* $F \nsubseteq F_2$.

Proof. Immediate from Part 6 of Proposition 36, together with the previous Corollary. ∎

Corollary 3. *If* F *is indecomposable, then* $F \nless \mathbf{C}$.

Proof. Suppose that $F < \mathbf{C}$. We will obtain a contradiction by defining a filter $F_0 > F$ with $F = \mathbf{C} \cap F_0$, as follows:

$$F_0 = \{W : (\exists V \in F)\ V_0 \leq W\}.$$

To see that F_0 is a filter, we consider $W, W' \in F_0$ and prove that $W \wedge W' \in F$. We have $W \geq V_0$ and $W' \geq V'_0$ with $V, V' \in F$. Thus, $W \wedge W' \geq V_0 \wedge V'_0 \geq (V \wedge V')_0$ (by part 3 of the Proposition). Thus, F_0 is a filter, and clearly, $F_0 \supseteq F$. We have assumed that F contains a variety $V < 1$, and hence, F_0 contains V_0 which, by Part 6, is not in the filter F_2. Since $F \subseteq \mathbf{C} \subseteq F_2$, we know $V_0 \notin F$. Thus, $F_0 > F$. Finally, we need to see that $F_0 \cap \mathbf{C} \subseteq F$ (the opposite inclusion being obvious). If $W \in F_0 \cap C$, then $W \geq V_0$ for some $V \in F$ and moreover $W \geq C$. And so by Part 5, $W \geq V_0 \vee C \geq V$, and so $W \in F$. ∎

In fact, a minor modification of the above proof will establish the next Corollary.

Corollary 4. *If* F *is a Mal'tsev filter and* $F < \mathbf{C}$, *then* F *is the proper intersection of two Mal'tsev filters.* ∎

The next two corollaries simply rephrase the results of Corollaries 2 and 3 for the case of principal filters.

Corollary 5. *If* $V \geq [x+y \approx y+x]$ *and* $V \nleq C$, *then* V *is not* $\vee$*-prime in* L. ∎

Corollary 6. *If* $C < V < 1$, *then* V *is not* $\vee$*-irreducible in* L. ∎

Thus, for instance, many familiar varieties fail to be $\vee$-irreducible: groups, rings, Boolean algebras, etc., with the examples too numerous to name. Among those that fail to be $\vee$-prime, we may mention semilattices and lattices. We do not know whether either of these varieties is $\vee$-irreducible.

Of course, the results of Proposition 36 and its Corollaries will also hold if we replace $[x+y \approx y+x]$ by $[Fxyz \approx Fyzx \approx Fyxz]$, and so on. It does not seem worthwhile to go into all the details.

<u>Problem</u>. *Is* V_0 $\mathbf{V}$*-irreducible for* V = *groups, rings, lattices, etc.?*

If in addition V is **V**-prime, then V must also satisfy either $V \leq C$ or $V \nleq [x+y \approx y+x]$. Clearly, *lattice theory* Lat has a commutative operation and cannot be defined from a single constant, hence Lat *is not* **V**-*prime*. We do not know whether it is **V**-irreducible.

7. k-TH ROOT FILTERS.

Though not strictly lattice-theoretical, there exists an interesting classification of filters on L, with the kth power construction playing much the same role that the join $\vee$ played for primality of filters. A filter F will be called a *k^{th} root filter* iff it satisfies the following condition:

$$V^{[k]} \in F \to V \in F.$$

Proposition 37. *The following conditions define k-th root filters for every k:*

congruence-modularity,

congruence-distributivity,

congruence-regularity,

congruence-weak-uniformity,

coherence,

congruence-permutability, and

spectrum $\subseteq$ J

for J *a submonoid of* $(\omega^+, \cdot, 1)$ *such that* $n^k \in J \to n \in J$.

Proof. Recall from Proposition 6 that if V is any variety, then the algebras of $V^{[k]}$ are k^{th} powers $\mathfrak{A}^k$ of algebras $\mathfrak{A} \in V$, with two extra operations

$$c(\langle\alpha_1,\ldots,\alpha_k\rangle) = \langle\alpha_k,\alpha_1,\ldots,\alpha_{k-1}\rangle$$

$$d(\langle\alpha_1^1,\ldots,\alpha_k^1\rangle,\ldots,\langle\alpha_1^k,\ldots,\alpha_k^k\rangle) = \langle\alpha_1^1,\ldots,\alpha_k^k\rangle.$$

One easily establishes the further fact that if $f : (\mathfrak{A}^k,c,d) \to (\mathfrak{B}^k,c,d)$ is any homomorphism of algebras in $V^{[k]}$, then f has the form

$$f(\langle\alpha_1,\ldots\alpha_k\rangle) = \langle g(\alpha_1),\ldots,g(\alpha_k)\rangle$$

for some homomorphism $g : \mathfrak{A} \to \mathfrak{B}$. From this, one easily checks that a congruence-block of ker f is a product of k blocks of ker g, and that Con$(\mathfrak{A}^k$, c,d) is isomorphic to Con $\mathfrak{A}$. From these and a few related results, one easily deduces all the assertions of the proposition.∎

Thus, for instance, if $J = \omega - \{2\}$, then K_J is prime but not a k-th root filter for any k. On the other hand, if $J = \{1,3,5,7,\ldots\}$, then K_J is a kth root filter for every k, but K_J is not prime.

In fact, if F is any filter on L, we may define

$$\sqrt[k]{F} = \{V : V^{[k]} \in F\}.$$

It is clear that $\sqrt[k]{F}$ is a filter, and that F *is a* k^{th} *root filter iff* $F = \sqrt[k]{F}$. We will see that in fact, if F is principal, then so is $\sqrt[k]{F}$. Thus, for an arbitrary variety W, we will define a variety $\sqrt[k]{W}$ as follows. For each n-ary W-operation F, we take k new operations $F_1,\ldots,F_k$, each of them nk-ary. And for each algebra $\mathcal{L} \in W$ with universe a kth power A^k, we define $\sqrt[k]{\mathcal{L}}$ to be the algebra with universe A and operations F_j defined as follows:

(*) $F(\langle a_1^1,\ldots,a_k^1\rangle,\ldots,\langle a_1^n,\ldots,a_k^n\rangle) = \langle F_1(a_1^1,\ldots,a_k^n),\ldots,F_k(a_1^1,\ldots,a_k^n)\rangle$.

One easily proves that the class of all algebras $\sqrt[k]{\mathcal{L}}$ forms a variety, which we will call $\sqrt[k]{W}$.

Proposition 38. (1) $V^{[k]} \geq W$ *iff* $V \geq \sqrt[k]{W}$, *thus if* F *is the principal filter generated by* W, *then* $\sqrt[k]{F}$ *is the principal filter generated by* $\sqrt[k]{W}$.

(2) *If* W *is a finitely based variety, then so is* $\sqrt[k]{W}$.

(3) $\sqrt[k]{V^{[k]}} \leq [V]$

(4) $(\sqrt[k]{V})^{[k]} \geq V$

(5) $V \geq W \leftrightarrow \sqrt[k]{V} \geq \sqrt[k]{W}$; *and thus* $\sqrt[k]{\ }$ *is well defined as an operation on* L.

Remark. The variety $\sqrt[k]{W}$, formula (*), etc., were introduced by R. McKenzie [56], and all of Proposition 38 is to be found in [56], either explicitly or implicitly. And so we will not give an entire proof of every detail.

Sketch of proof of Proposition 38. It is helpful to think in terms of Proposition 1 of the Introduction. For instance, for (1), suppose we had a universe-preserving functor $\Phi : V \to^{[k]} W$. Then, given a model of $V^{[k]}$, is naturally isomorphic (see Propositions 6 and 7) to an algebra whose V-reduct is a kth power $\mathfrak{A}$. Applying Φ to $\mathfrak{A}$ and then allowing (*) to define the operations F, we obtain a universe-preserving functor $\Psi : V^k \to W$. Conversely, if we are given Ψ and wish to construct Φ, we may take an algebra $\mathfrak{A} \in V$, take its kth power $\mathfrak{A}^k$ and then add operations c and d according to the formulas in the proof of Proposition 37. If we then apply Ψ and use (*) to define operations F_j, we obtain our functor Φ.

The reasoning for (3) is similar and omitted. The reader may be interested to try using (*) to suggest the defining identities for $\sqrt[k]{W}$; details may be found in [56]. See also the example after Proposition 39.

Finally, (4) is an immediate Corollary of (1), and (5) is an immediate Corollary of (1) and (4). ∎

Remark. Of course, a strictly formal proof of parts (1) and (3) may be given. One may use the semantic argument just given to *suggest* the correct interpreting terms. We omit details.

Corollary. *If* F *is the principal filter generated by* W, *then* F *is a* k^{th} *-root filter iff* $W = \sqrt[k]{W}$ *in* L. ∎

Of course, this condition will hold if $W^{[k]} \leq W$ (like the example in Proposition 7; see also the remarks after the Corollary to Proposition 11). But there are some other surprising examples, namely those varieties which are defined by *simple laws*. A *simple term* (in this context) will mean either a variable or a term $F(\alpha_1,\ldots,\alpha_n)$ where F is a function symbol and each α_i is a variable. A *simple identity* is an identity $\sigma \approx \tau$ with σ and τ simple terms. The following Proposition was known to R.N. McKenzie (unpublished), and to B. Davey and H. Werner; see p. 197 of [17].

Proposition 39. *If* W *is definable by simple identities, then* $W = \sqrt[k]{W}$ *for every* k.

Proof. For $\mathfrak{B} \in \sqrt[k]{W}$, B^k has the structure of a W-algebra. Now for any n-ary W-operation F on B^k we define an n-ary operation F' on B as follows:

$$B^n \xrightarrow{\Delta} (B^k)^n \xrightarrow{F} B^k \xrightarrow{\pi_1} B$$

(the composite $B^n \to B$ being F')

where π_1 is first-coordinate projection, F is as given and $\Delta(b_1,\ldots,b_n) = (b_1,\ldots,b_1,b_2,\ldots,b_2,\ldots,b_n,\ldots,b_n)$. Let us investigate the operation defined on B by the m-ary simple term $\beta = F'(x_{i_1},\ldots,x_{i_n})$ where $(i_1,\ldots,i_n)$ is a sequence selected from $\{1,\ldots,m\}$. This operation is given by

$$B^m \xrightarrow{\lambda} B^n \xrightarrow{\Delta} (B^k)^n \xrightarrow{F} B^k \xrightarrow{\pi_1} B$$

where $[\lambda(\alpha)]_j = \alpha_{i_j}$ $(1 \le j \le n)$. But the operation $G(x_1,\ldots,x_m) = F(x_{i_1},\ldots,x_{i_n})$ leads to

$$G' = B^m \xrightarrow{\Delta} (B^k)^m \xrightarrow{G} B^k \xrightarrow{\pi_1} B$$

$$= B^m \xrightarrow{\Delta} (B^k)^m \xrightarrow{\mu} (B^k)^n \xrightarrow{F} B^k \xrightarrow{\pi_1} B.$$

where $[\beta(\gamma)]_j = \gamma_{i_j}$ $(1 \le j \le N)$ (with each $\gamma_i \in B^k$ $(1 \le i \le M)$). One easily checks that $\mu\Delta = \Delta\lambda$, and hence that G' is defined by β on B. Now it follows that the defined operations F' obey every simple identity which holds for the operations F. ∎

Of course the same remark applies here as to the proof of Proposition 38. Here we will illustrate the formal procedure with an example. The Mal'tsev identities are the following:

(Mt_1): $pxzz \approx x;\quad pxxz \approx z.$

If we apply (*) to these identities, we obtain the following defining identities for $\sqrt[2]{Mt_1}$:

$$p_1(x,u,z,w,z,w) \approx x$$

$$p_2(x,u,z,w,z,w) \approx u$$

$$p_1(x,u,x,u,z,w) \approx z$$

$$p_2(x,u,x,u,z,w) \approx w.$$

To interpret Mt_1 in $\sqrt[2]{Mt_1}$ we clearly may define

$$pxyz = p_1(x,x,y,y,z,z).$$

[Doubling each variable corresponds precisely to the map Δ in the previous proof, and disregarding p_2 corresponds to the map π_1 in that proof.]

Thus, for instance, the theory $Sem\ell$ of semilattices has $\sqrt[k]{Sem\ell} = Sem\ell$ in L, even though (by Proposition 7) $Sem\ell < Sem\ell^{[k]}$. By contrast, $\sqrt[2]{G} < G$ for G the variety of groups. Referring to Proposition 8 and the definition of File just before Proposition 8, we note an interesting and surprising way to prove that $\sqrt[2]{G} < G$: there exists a topological space A which can model an algebra in $\sqrt[2]{G}$ but cannot model an algebra in G. (Thus we have $\mathcal{A} = (A; F_1, F_2, \ldots) \in \sqrt[2]{G}$ with the topology of A in File $\mathcal{A}$ but not in File $\mathcal{L}$ for $\mathcal{L}$ any group.) In other words, A^2 can be a topological group but A itself cannot. For the existence of such a space A, we refer the reader to the work of J. van Mill [59], who has found a space A which has *no* autohomeomorphism (and hence cannot be a topological group), but such that A^2 is homeomorphic to the Hilbert cube. (See also [5], [31], [50], and [65] for some surprising "roots" of topological spaces.) But, for instance, we note that by Proposition 38 (or by the remarks just above), van Mill's space must admit a continuous Mal'tsev term. (Thus it is in fact an H-space; see [78].)

Perhaps the van Mill example will help the reader appreciate the basic simplicity of the sometimes confusing technicalities surrounding the notions of $V^{[k]}$ and $\sqrt[k]{W}$. One may simply remember, e.g., that there exists a space A such that A^2 supports a topological group but A does not; by contrast, no such example is possible for, say, semilattices. Then the operations of "$\sqrt[2]{G}$" are simply those operations which *must* be present on a space A if we are to obtain a group structure upon squaring A.

8. TENSOR PRODUCTS ON L.

Our lattice has another interesting structural feature which is not strictly lattice-theoretical, namely the *tensor product*. Returning to the notation established just before Proposition 3, we will say that the *tensor product variety* $V \otimes W$ has precisely the same operations as the coproduct variety $V \coprod W$ and is axiomatized by the coproduct identities $\Sigma \cup \Gamma$ together with all identities

$$F_s(F_t(x_{11},x_{12},\ldots),F_t(x_{21},x_{22},\ldots),\ldots)$$

$$\simeq F_t(F_s(x_{11},x_{21},\ldots),F_s(x_{12},x_{22},\ldots),\ldots)$$

with $s \in S$ and $t \in T$. Thus, $V \otimes W = W \otimes V$ in L (in fact, they are obviously equivalent). From the categorical point of view, $V \otimes W$ is the category of all models of V in the category W (or vice versa). Since $V \otimes W$ is a subvariety of $V \coprod W$ we obviously have $V \otimes W \geq V \vee W$. Equality holds if $V = 0$ in L, but generally $V \otimes W$ is much greater than $V \vee W$, as we shall see.

Proposition 40. (1) *If* $V \leq U$ *then* $V \otimes W \leq U \otimes W$, *and hence* $V \otimes W$ *is well defined as an operation on* $L \times L$.

(2) $V \otimes (U \otimes W) = (V \otimes U) \otimes W$ *in* L.

(3) $V \otimes (U \wedge W) = (V \otimes U) \wedge (V \otimes W)$ *in* L.

Proof. (1) is obvious. For (2), each side may be thought of as the subvariety of $U \coprod V \coprod W$ in which U-operations commute with V-operations, V-operations commute with W-operations and W-operations commute with U-operations. For (3), the inequality $V \otimes (U \wedge W) \leq (V \otimes U) \wedge (V \otimes W)$ is immediate from (1). For the opposite inequality, we will create a universe-preserving functor $V \otimes (U \times W) \to (V \otimes U) \times (V \otimes W)$, as follows. An algebra $\mathfrak{C} \in V \otimes (U \times W)$ is a model of $U \times W$ in the category V. By Proposition 5, $\mathfrak{C}$ is isomorphic to $\mathfrak{A} \times \mathfrak{B}$ with $\mathfrak{A}$ a model of U in V and $\mathfrak{B}$ a model of W in V. This immediately yields an algebra in $(V \otimes U) \times (V \otimes W)$. We omit further details concerning this functor. ∎

Now for any $V \in L$, there is defined a filter

$$\text{In } V = \{W : W \otimes V = 1\}.$$

Among proper such filters there is a largest one as was proved by J. Isbell [41].

Proposition 41. *If* V < 1, *then* In V $\subseteq$ In V_2. ∎

Isbell also asked whether there exist $2^{2^{\aleph_0}}$ distinct filters of the form

$$\bigcap_{V \in K} \text{In } V$$

for some class K of varieties. This remains unknown. Notice that if V is a finitely based variety of finite type, then In V is a Mal'tsev filter (see §5). Moreover, a better description (e.g., an (explicit) Mal'tsev condition) for In V_2 remains unknown. Nevertheless, a recent result of Csákány will show us that In V_2 is prime, and give us an easily described class $\{W_\kappa : \kappa$ any ordinal$\}$ of varieties generating the prime ideal complementary to In V_2.

For each infinite cardinal κ, we define W_κ to be the variety $V(\kappa)$ in the *dual* of the category of sets (using the notation at the beginning of §2). It is not difficult to see that if V < 1, then $W_\kappa \notin$ In V, as follows. Given an algebra $\mathfrak{A} = (A;F_t)_{t \in T} \in V$, we obtain a model of W_κ in the category V, as follows. Consider the product-preserving functor $\text{Sets}^{\text{Dual}} \to \text{Sets}$ which takes an arbitrary $F : X \to Y$ to the "dual" map $A^F : A^Y \to A^X$, defined by $A^F(K) = F^{-1}[K]$. Thus, for any (dual) operation $F : \kappa \to n\kappa$ of W_κ, we have the corresponding operation $A^F : (A^\kappa)^n = A^{n\kappa} \to A^\kappa$, which is obviously a homomorphism of V-algebras. Thus

$$\mathfrak{L} = (A^\kappa; F_t, \ldots F \ldots)_{t \in T}, \quad F \text{ as above}$$

is a non-trivial algebra in $V \otimes W_\kappa$. Thus $V \otimes W_\kappa < 1$, and so $W_\kappa \notin$ In V.

Now Csákány's recent result is that if $\mathfrak{L}_0$ is any non-trivial algebra in $V_2 \otimes W$ (for any W), then $\mathfrak{L}_0$ can be represented as a subalgebra of a reduct of an algebra $\mathfrak{L} = (A^\kappa; F_t, \ldots F \ldots)$ as above, with $(A;F_t) \in V_2$. Moreover, by taking powers, we may assume $|A|$ is large. Thus, such a $\mathfrak{L}$ also naturally has the structure of a $(V \otimes W)$-algebra for any V modelable on the universe A, i.e., any V < 1. And so we obtain, as a Corollary of Csákány's result, an immediate reproof of Isbell's theorem stated above. Moreover, this result gives a structural description for a standard form of an algebra in $V_2 \otimes W < 1$, yielding a proof of the next Corollary (in the manner of the various proofs in §5).

Corollary (of Csákány [16]). *In* V_2 *is a prime filter.* ∎

Csákány has called *ubiquitous* those varieties which are not in In V_2. They, of course, form a prime ideal, which we will denote Ub.

Corollary (of Csákany [16]). Ub *is the ideal generated by all* W_κ (for κ any cardinal). ∎

Finally we will prove the following Proposition, which shows that the ideal Ub is unbounded in L.

Proposition 42. $\bigvee_{\kappa \in \mathrm{Card}} W_\kappa = 1$ *in* L.

Proof. We will prove that if $V < 1$, then $W_\kappa \not\leq V$ for $2^\kappa >$ the number of operations of V. For each $D \subseteq \kappa$, the object κ in the category of sets has the dual operation

$$A \xrightarrow{F_D} A \times \{0\} \cup A \times \{1\}$$

defined by

$$F_D(a) = (a, \varepsilon_D(a))$$

where

$$\varepsilon_D(a) = \begin{cases} 0 & a \in D \\ 1 & a \notin D. \end{cases}$$

Now, for $D, E \subseteq \kappa$ with $|D - E| = \kappa$, there exists a one-one map $q_{DE} : \kappa \to \kappa$ with range $\subseteq D - E$. Let p_{DE} be a left-inverse: $p_{DE} \circ q_{DE} = 1$. Now as operations in a variety, these maps obey the following laws:

$$F_D(x,x) \approx x$$

$$q_{DE}F_D(x,y) \approx q_{DE}(x)$$

$$q_{DE}F_E(x,y) \approx q_{DE}(y)$$

$$q_{DE}(p_{DE}(x)) \approx x.$$

Now it is clear that for such D and E, F_D and F_E must have distinct images under any interpretation of W_κ in V. Otherwise, denoting $F_D = F_E$ by F, p_{DE} by p and q_{DE} by q, we would have:

$$x \approx qpx \approx qF(px,py)$$

$$\approx qpy \approx y$$

as identities in V, in contradiction to our assumption that $V < 1$. The Proposition now follows by taking 2^κ subsets D, all differing on subsets of power κ. ∎

We close this section by recalling two results from [76]:

(a) $\mathrm{Sem}\ell \otimes \mathcal{G} = 1$

(b) If $V > 0$ and V is an idempotent variety, then $V \otimes \mathcal{G} \geq \mathrm{Ab}$.

($\mathcal{G}$ = groups, Ab = Abelian groups).

9. LOCATION OF VARIETIES IN L WITH EMPHASIS ON VARIETIES OF GROUPS.

As mentioned in the introduction, we first obtain a rough idea about L, described by Figure 1 (located on page 7), and immediately focus on varieties of groups, whose results we collect in Figures 2, 3 and 4. An effort has been made to keep all these pictures readable. The results in the discussions contain more detail, and surely will help one to properly understand such pictures.

Both for this and for the next section, we refer the reader to the definitions of File( ) and SIN algebras and to Proposition 8 and Proposition 9 near the end of section 1.

Throughout the next sections, we shall apply Proposition 1 very frequently. We shall refer to functors which commute with the underlying set functors as "concrete functors."

Before the first section after the figures which we include, is a key for the notation used for varieties. Both this key and Figs. 1 through 7, are also useful in §10 and §11.

One should be aware that most of the propositions that follow also have alternative proofs, as there are reasons of various kinds for these results. See, for example, the alternative proof of Proposition 47 below.

Before we introduce the first proposition of this chapter, we recall some of the notation of our key: Ab, Ab_n, An, An_k, Anu_k, Ancu, $Ancu_n$, Ancup, and will denote respectively the varieties of abelian groups, abelian groups of exponent n, rings, rings which satisfy $kx = 0$, rings with 1 of characteristic k, commutative rings with 1, commutative rings with 1 of characteristic n, and the subvariety of $Ancu_p$ with the additional equation $x^p \approx x$. By V_n we understand the variety of Post algebras generated by an n-element Post algebra as introduced in §2.

<u>Proposition 43</u>.

i) $Ab = {}_L An$

ii) $Ab_k = {}_L An_k$ *for all* $k > 1$

iii) $Ab_k < Anu_k \leq Ancu_k$ *for all* $k > 1$

iv) $An < Ancu < V_{\aleph_0}$

v) $\text{Ancu}_n \leq \text{Ancun} \leq V_n$ *for* n *prime*

vi) $\text{Ancu}_n < V_n$ *for* n *non-prime*

vii) $\text{Ancu}_m \leq \text{Ancu}_n$ *if* $n|m$

viii) $\text{Ancu}_{pn} < \text{Ancu}_n$ *if* p *is prime and* $p|n$

Proof. The forgetful functor from rings into abelian groups (forgetting multiplication, and 1 if it exists) provides immediately $\text{Ab} \leq \text{An}$, $\text{Ab}_k \leq \text{An}_k \leq \text{Anu}_k$. Since every abelian group can be interpreted on a ring by defining the multiplication "$a \cdot b = 0$" we obtain i) and ii). Suppose now that $\text{Anu}_k \leq \text{Ab}_k$, which provides us with $F : \text{Ab}_k \to \text{Anu}_k$ a concrete functor. This means that there are terms $qx + \ell y$ and $nx + my$ in $F_2\text{Ab}_k$ which define the ring operations in Anu_k : $a + b = qa + \ell b$ and $a * b = na + mb$.

Then $0 + 0 = q0 + \ell 0 = 0$ and the 0 of every abelian group in Ab_k is also the 0 of its interpretation in Anu_k. Then $0 = 0 * a = ma$ and $0 = a * 0 = na$. Since $\mathbb{Z}_k \in \text{Ab}_k$ in particular one has $0 = m \cdot 1 = m$ and $0 = n \cdot 1 = n$. The multiplication in $F(\mathbb{Z}_k)$ is the $a * b = 0$, and a contradiction is obtained since an element of $F(\mathbb{Z}_k)$ will be its 1, and $1 * 1 = 0$ provides $1 = 0$ leading to a ring with 1 element only. Since $\text{Ancu}_k \subseteq \text{Anu}_k$ is a subvariety, $\text{Anu}_k \leq \text{Ancu}_k$ follows. The same argument for $\text{Ab}_k < \text{Anu}_k$ also shows that $\text{Ab} < \text{Ancu}$. Since $(\mathbb{Z},+,-,\cdot,0,1) \in \text{Ancu}$ is a model of cardinality $\aleph_0$, $\text{Ancu} \leq V_{\aleph_0}$. However, $V_{\aleph_0} \nleq \text{Ancu}$, since $(\mathbb{Z}_2,+,-,\cdot,0,1) \in \text{Ancu}$ and $2 \notin \text{Spec } V_{\aleph_0}$. The first part of v) is obvious as Ancun is a subvariety of Ancu_n. For the second part we use the well known fact that the fields $\mathbb{Z}_p$ are functionally complete, i.e., every n-ary function is a polynomial. (One of the standard arguments is that there are p^{p^n} different polynomials, also different when considered as n-ary functions and this is precisely the number of possible n-ary functions). The varieties Ancup are now $\text{ISP}(\mathbb{Z}_p) =_L Vp$.

To show vi) let n be non-prime and p a prime such that $p|n$. Since the ring $\mathbb{Z}_n$ is in Ancu_n, and n-element model, $\text{Ancu}_n \leq V_n$. But $\mathbb{Z}_p$ is also a ring in Ancu_n and $p \notin \text{Spec } V_n$. Therefore $V_n \leq \text{Ancu}_n$. If $n|m$, it is clear that Ancu_n Ancu_m, as $n \cdot 1 = 0$ implies that $m \cdot 1 = 0$, and vii) follows. Finally, if p is a prime such that $p|n$ by vii) just shown, one has $\text{Ancu}_{pn} \nleq \text{Ancu}_n$ and since $p \in \text{Spec}(\text{Ancu}_{pn})$ but $p \notin \text{Spec}(\text{Ancu}_n)$, $\text{Ancu}_n \nleq \text{Ancu}_{pn}$ follows. So $\text{Ancu}_{pn} < \text{Ancu}_n$. ∎

The following algebras are of general use in computers: On a set $S_n := \{z \in \mathbb{Z} \,|\, 0 \leq z \leq n\}$ one defines $+_n$ to coincide with the usual sum except when such would give a result larger than n. Then the $+_n$ addition gives n. A $\cdot_n$ multiplication is defined in a similar way and the "integral division" /

gives the largest integer smaller than what usual division would give. For 0/0 we set 1.

The algebra S_n is functionally complete: $p_f(x_1,\dots,x_k)$

$$= \sum_{(i_1,\dots,i_k)\varepsilon S^k} f(i_1,\dots,i_k)(x_1/i_1\cdot i_1/x_1\cdot x_2/i_2\cdot i_2/x_2\cdot\ldots\cdot x_k/i_k\cdot i_k/x_k)$$

is such that $p_f(x_1,\dots,x_k) = f(x_1,\dots,x_k)$ for any k-ary f. The above discussion has the following

Corollary.

$$\mathrm{HSP}S_n =_L V_n.\blacksquare$$

Thus the apparently very abstract, or "artificial" varieties V_n are in fact equal in L to varieties generated by the today most widely used algebras.

Proposition 44.

$$\bigvee_{\alpha\ \varepsilon\ \text{cardinal}} V_\alpha = 1 \quad .$$

Proof. Let $V \neq 1$. Then V has a model M such that $1 < |M| = \lambda$. Therefore, $V \leq V_\lambda < V_{2\lambda}$. Thus, 1 is the least upper bound of the class $(V_\alpha)_{\alpha\ \varepsilon\ \text{Card}}$ in L.$\blacksquare$

Proposition 45. *Let* p *be a prime number. Then,*

$$\mathcal{A}ncu_p < \mathcal{A}ncup.$$

Proof. Since $\leq$ is clear, we suppose that $\mathcal{A}ncup \leq \mathcal{A}ncu_p$, and look for a contradiction. Let $t(x,y), u(x,y)\ \varepsilon\ F_2\mathcal{A}ncu_p$, be respectively the interpretations of $\mathcal{A}ncu_p$-multiplication and addition. Consider $t(x,y)$. Since $t(x,y) = t(y,x)$,

$$t(x,y) = a_0 + a_1(x+y) + a_2xy + a_3(x^2+y^2) + a_4(x^2y+xy^2) + a_5(x^3+y^3) + a_6x^2y^2 + a_7(x^3y+xy^3) + a_8(x^4+y^4) + \ldots + a_k(x^ny^m+x^my^n)$$

and we can assume without loss of generality $n \geq m$.

Suppose $n > 1$, and consider the associativity of t. Now

$$t(t(x,y),z) = \ldots\ldots + a_k(t(x,y)^nz^m + t(x,y)^mz^n)$$

has a coefficient for $(x^ny^m)^nz^m$, a_k^{n+1}. However, the coefficient of the same monomial in $t(x,t(y,z))$ is zero as the monomial here has a higher power of x than x^n.

Thus $a_k = 0$. We now delete a_k and the same argument shows $a_{k-1} = 0$ until $a_3 = 0$. Therefore, $t(x,y) = a_0 + a_1(x+y) + a_2xy$. Similarly, $u(x,y) = b_0 + b_1(x+y) + b_2xy$.

Define: $x^{1(t)} = x$ and $x^{n+1(t)} = t(x,x^{n(t)})$

Then $x^{p(t)} = a_0 + \dots + a_2^{p-1}x^p$

Since $x^{p(t)} = x$, $a_2^{p-1} = 0 \therefore a_2 = 0$. Similarly, since

$$x^{p(u)} = 0_u \text{ a constant, } b_2^{p-1} = 0 \therefore b_2 = 0$$

From the associativity $a_1^2 = a_1$ and $b_1^2 = b_1$. Since constant operations are not ring-operations, $a_1 = b_1 = 1$. Therefore

$$t(x,y) = a_0 + x + y$$

$$u(x,y) = b_0 + x + y.$$

Since $-b_0 = u(-b_0,-b_0)$, $-b_0 = 0_u$

$$-b = 0_u = t(0_u,0) = a_0 + (-b_0) + 0 \therefore a_0 = 0$$

$$0_u = t(0_u,0_u) = 0_u + 0_u \qquad 0 = 0_u = -b_0 \therefore b_0 = 0$$

Therefore, $t(x,y) = x + y = u(x,y)$.

This leads to $0_u = 1_t$, and so the ring interpreted would have only one element. There are however non-trivial rings in $Ancu_p$, like $\mathbb{Z}_p$. ∎

Proposition 46. *If* Lat *is the variety of lattices, then* Lat $<$ V_n *for every natural number* n.

Proof. The spectrum argument yields Lat $\leq V_n$ immediately, since Spec(Lat) = $\mathbb{N}$. And $V_n \nleq$ Lat is also obvious for same reason. ∎

As shown in Proposition 43, Ancu3 $=_L V_3$ and from Proposition 4, Lat $< V_3$. One could also give the explicit terms in most of these cases. For example, the following two derived operations of Ancu3

$$a \wedge b = ab + 2a^2b + 2ab^2 + 2a^2b^2$$

$$a \vee b = a + b + 2ab + a^2b + ab^2 + a^2b^2$$

provide a concrete functor F : Ancu3 $\to$ Lat and Lat $<$ Ancu3 follows.

The argument used in Proposition 46 also yields Ab $<$ Ab_n for all natural n, and if N_k means the variety of vector spaces over a field k of infinite cardinality, one obtains Ab $< N_k$ and $N_k < V$ if $\mu > |k|$. The varieties N_k can also be used to show that L is a proper class.

In the following proposition Semℓ denotes the variety of semilattices.

Proposition 47.

$$\mathrm{Sem}\ell < \mathrm{Lat}.$$

Proof. Semℓ $\leq$ Lat is obvious by considering the appropriate forgetful functor. Let B : = , a three-element semilattice. Since the only lattice A with three elements  is rigid, File(B) $\not\leq$ File(A), and therefore Lat $\not\leq$ Semℓ.

Alternatively we know that Con(A) is distributive for any lattice, but there exists a semilattice B with Con(B) not distributive, e.g.  = : B, and so again File(B) $\not\leq$ File(A). ∎

We do not discuss here the rest of Fig. 1 as the rest is precisely the subject of §10.

For our next Proposition we shall use the following lemma of [66]:

Lemma. *Let* V *and* W *be varieties of groups. Let* $H : V \to W$ *be a concrete functor. If* $V \subseteq \mathrm{Ab}$, *then* H *is the inclusion functor.* ∎

This means: *If* $V \subseteq \mathrm{Ab}$ *and* $W \leq V$, *then* V *is a subvariety of* W.

Proposition 48. *If* V *and* W *are varieties of abelian groups and* V *is a proper subvariety of* W, *then* $W < V$. ∎

Proposition 49. *Let* $(\mathbb{N}, |)$ *be the lattice of natural numbers with divisibility as a partial order. Then* $\mathrm{Ab}_m \wedge \mathrm{Ab}_n = \mathrm{Ab}_{m \vee n}$ *for all* $m, n \in \mathbb{N}$.

Proof. We first show that if p, q are different primes and $k, \ell \in \mathbb{N}$, then $\mathrm{Ab}_{p^k} \wedge \mathrm{Ab}_{q^\ell} =_L \mathrm{Ab}_{p^k q^\ell}$.

For that purpose, define a functor

$$F : \mathrm{Ab}_{p^k q^\ell} \to \mathrm{Ab}_{p^k} \wedge \mathrm{Ab}_{q^\ell}$$

as follows: Let $(G, \cdot, i, 1) \in \mathrm{Ab}_{p^k p^\ell}$. Then $G = G_p \times G_q$ where G_π is the π-Sylow subgroup of G.

Considering the definition of "$\wedge$ in L" in section 1, Proposition 3, we define

$$F_p(G_p) := (G_p, \cdot, i, 1, \otimes, \overset{x}{i}, \overset{x}{1}, \overset{0}{\ }) \in \mathrm{Ab}_{p^k}$$

where $(G_p, \cdot, i, 1) \in \mathrm{Ab}_{p^k}$ is the p-Sylow subgroup of G, and $x \otimes y = x$, $\overset{x}{i}(x) = x$, $\overset{x}{1}(x) = x$, $\pi(x,y) = x$.

Similarly $F_q(G_q) := (G_q,\cdot,\mathring{i},\mathring{1},\cdot,i,1,\pi) \in Ab^0_{q^\ell}$ where $(G_q,\cdot,i,1) \in Ab_{q^\ell}$ is the q-Sylow subgroup of G and $x \circ y = x$, $\mathring{i}(x) = x$, $\mathring{1}(x) = x$, $\pi(x,y) = y$.

We define $F(G) := F_p(G_p) \times F_q(G_q)$. If $h : G \to H$ is an $Ab_{p^k p^\ell}$-homomorphism, since $h(G_p) \subseteq H_p$ and $h(G_q) \subseteq H_q$, $h = (h_p,h_q)$ with $h_\gamma = h/G_\gamma$. Clearly $F_p(h_p) : F_p(G_p) \to F_p(H_p)$ is a homomorphism, as $\{\otimes,\overset{x}{i},\overset{x}{1},\pi\}$ are projections. Similarly for $F_q(h_q)$. We define $F(h) := (F_p(h_p),F_q(h_q)) : F(G) \to F(H)$, and so complete the concrete functor F.

Having proved our first claim, an obvious generalization will yield:

If $n = p_1^{n_1} \ldots p_k^{n_k}$, then $Ab_n =_L Ab_{p_1^{n_1}} \wedge \ldots \wedge Ab_{p_k^{n_k}}$.

Now let $m,n \in \mathbb{N}$ and $m = p^{m_1} \ldots p_k^{m_k}$ and $n = p_1^{n_1} \ldots p_k^{n_k}$.
Then

$$\begin{aligned} Ab_m \wedge Ab_n &= (Ab_{p_1^{m_1}} \wedge \ldots \wedge Ab_{p_k^{m_k}}) \wedge (Ab_{p_1^{n_1}} \wedge \ldots \wedge Ab_{p_k^{n_k}}) \\ &= (Ab_{p_1^{m_1}} \wedge Ab_{p_1^{n_1}}) \wedge \ldots \wedge (Ab_{p_k^{m_k}} \wedge Ab_{p_k^{n_k}}) \\ &= Ab_{p_1^{\max\{m_1,n_1\}}} \wedge \ldots \wedge Ab_{p_k^{\max\{m_k,n_k\}}} \\ &= Ab_{m \vee n}\,. \blacksquare \end{aligned}$$

In the proof of Proposition 49, π is the operation denoted by juxtaposition in the definition just before Proposition 3. In this case, the operation π of F(G) is a derived operation of G which we can explicitly describe. Since p^k and q^ℓ are relatively prime natural numbers, there are two integers s and t such that $sp^k + tq^\ell = 1$. One has $\pi(x,y) = x^{tq^\ell}y^{sp^\ell}$.

We now recall our definition of SIN algebras given after Proposition 8: A group G is a SIN group if all interpretations of G in Gru are isomorphic groups.

<u>Proposition 50</u>. *Let* $\mathbb{Z}_n$ *denote the cyclic group of order* n. *Then* $\mathbb{Z}_n$ *is* SIN *in* Gru.

<u>Proof</u>. Suppose V and W are subvarieties of Gru and $\mathbb{Z}_n \in W$, and let $F : W \to V$ be a concrete functor. Therefore, there exists a term $\alpha(x,y) \in F_2W$ such that for every algebra $\mathfrak{A} = (A,\cdot) \in W$ one has $F\mathfrak{A} = (A,\alpha) \in V$.

Since $(\mathbb{Z}_n,+_n) \in W$, then $(\mathbb{Z}_n,\alpha_{\mathbb{Z}_n}) \in V$. Let $\alpha(x,y) = x^{n_1}y^{m_1}x^{n_2}y^{m_2}\ldots$.
Let $M = \sum m_i$, $N = \sum n_i$. Then $0\alpha_{\mathbb{Z}_n}1 = M$, $1\alpha_{\mathbb{Z}_n}0 = N$ and in general

$$x\,\alpha_{\mathbb{Z}_n}\,y = Nx +_n My.$$

Associativity requires $N = N^2$ and $M = M^2$. Therefore $N = 0$ or 1 and $M = 0$ or 1. Moreover $0\alpha_{\mathbb{Z}}0 = 0$, and since $(\mathbb{Z}_n, \alpha_{\mathbb{Z}_n})$ is a group, 0 is the identity of it.

Now $x\alpha_{\mathbb{Z}_n}0 = x$. Therefore $N = 1\alpha_{\mathbb{Z}_n}0 = 1$ and $0_{\mathbb{Z}_n}y = y$. Therefore $M = 0\alpha_{\mathbb{Z}_n}1 = 1$. This says that $x\alpha_{\mathbb{Z}_n}y = x +_n y$ and $\alpha_{\mathbb{Z}_n} = +_n$. ∎

As in [40] we understand by the "*Order Structure*" of a finite group G of order n the n-tuple

$$(0_{G,1}, 0_{G,2}, \ldots, 0_{G,n})$$

where $0_{G,i}$ means the number of its elements of order i.

Lemma. *Let K, M be subvarieties of* Gru *and* $F : K \to M$ *a concrete functor. Then*

$$0_{G,m} \leq 0_{F(G),m} \quad \textit{for all } m.$$

Proof. The number of different subgroups of G which are isomorphic to $\mathbb{Z}_m$ is $0_{G,m}/0_{\mathbb{Z}_m,m}$. Since by Proposition 8, $F(G)$ has at least as many subgroups isomorphic to $\mathbb{Z}_m$, one has

$$0_{G,m}/0_{\mathbb{Z}_m,m} \leq 0_{F(G),m}/0_{\mathbb{Z}_m,m}$$

and the result follows. ∎

Our next result simplifies very much the task of finding adequate SIN groups which allow the proofs of several of the subsequent propositions.

Proposition 51. *Let G be a finite group. Let K, M be varieties of groups such that $G \in K$. Then for every concrete functor $F : K \to M$,*

$$(0_{G,1}, 0_{G,2}, \ldots, 0_{G,|G|}) = (0_{F(G),1}, 0_{F(G),2}, \ldots, 0_{F(G),|G|})$$

Accordingly F preserves the exponent of G.

Proof.

$$|G| = \sum_{i=1}^{|G|} 0_{G,i} \leq \sum_{i=1}^{|F(G)|} 0_{F(G),i} = |F(G)| = |G|.$$

Since by the lemma $0 \leq 0_{G,i} \leq 0_{F(G),i}$ for all i, the result follows. ∎

We already commented about this Proposition in §1 when we provided examples of SIN algebras.

Before we proceed with the varieties of groups, and for convenience of reference, we recall the following definition from Lattice Theory.

Definition. Let $(L,\leq)$ be a lattice and $S \subseteq L$. We call S a chain if $a < b$ or $b < a$ for any two different elements a,b of S. We call S an antichain if $a \nleq b$ and $b \nleq a$ (in short "$a \| b$") for any two different elements of S. We call h the height of an element a of S if h is the cardinality of the shortest maximal chain (which cannot be refined by inserting elements at the beginning or in between) $a_1 < a_2 < \dots < a_n = a$. We call b the breadth of S if b is the supremum of cardinalities of antichains in S.

Proposition 52. *For every natural number* n *the breadth of the* L-*interval* $[Ab, Ab_n]$ *is infinite.*

Proof. $\{Ab_{pn} \mid p$ prime and $p \nmid n\}$ is a subset of this interval, and since $\mathbb{Z}_{pn}$ and $\mathbb{Z}_{qn}$ are SIN groups, by Proposition 51, $Ab_{p^n} \| Ab_{q^n}$ for $p \neq q$. ∎

Proposition 53. *Let* B_n *denote the Burnside variety of groups of exponent* n. *Let* m,n *be natural numbers,* $m \neq n$ *and* $m|n$. *Then*

i) $B_n < B_m$

ii) $[Gru, B_n]$ has infinite breadth.

iii) $B_n \cap X < B_m \cap X$ for any variety X of groups defined by commutators.

Proof. For i) and iii) the variety on the right is a subvariety of the variety of the left. Therefore "$\leq$" is obvious. Since $\mathbb{Z}_n \in B_n \cap X \subseteq B_n$ and it is a SIN group, by Proposition 51 a concrete functor $F : B_m \to B_n$ or $F : B_m \cap X \to B_n \cap X$ is not possible. For ii) we consider $\{B_{pn} \mid p$ prime and $p|n\}$ a subset of $[Gru, B_n]$ and show that it is an antichain by using Proposition 51 exactly as we did in Proposition 52. ∎

We thank J.A. de la Peña [66] for having provided us a better proof of our original statement. Our Proposition 51, however, now provides an even clearer and simpler argument.

To continue our study of varieties of groups, we first call the reader's attention to Theorem 12.12 of [62] according to which *every law of groups is equivalent to a pair of laws, one of the form* $x^m \approx 1$ *and the other* $c \approx 1$ *with* c *a commutator*. This well known result due to B.H. Neumann (1937) will explain our choices for first comparisons of varieties of groups.

Our notation will often approach that of [62], but we prefer to be explicit.

We shall denote by N_n the variety of groups defined by the equation $[x_1, x_2, \dots, x_{n+1}] \approx 1$, where the square brackets mean "commutators," and $[x_1, \dots, x_n, x_{n+1}]$ means $[[x_1, \dots, x_n], x_{n+1}]$. We call the groups in N_n n-*nil-*

potent groups. By M we denote the variety of groups defined by the equation $[[x,y],[z,w]] \approx 1$, which we call *metabelian*, and by M_n we mean the variety of models of $[v,w] \approx 1$ where $M_1 = M$ and M_{n-1} is the variety of models of $v \approx 1$ or of $w \approx 1$.

To simplify our notation, we shall write $B_n N_m$ rather than $B_n \cap N_m$ and delete the "$\cap$" signs in this manner. So $B_n N_m M$ or $B_n M N_m$ will denote $B_n \cap N_m \cap M$.

Among our examples of SIN algebras we have mentioned in §1 the dihedral groups. We also remarked that this example would be useful for Proposition 54.

Lemma. *Every dihedral group is* SIN *in* Gru.

Proof. By Proposition 51 it is sufficient to show that if G is a group with the same order structure as D_n, then $G \cong D_n$.

Now the order structure of D_n is as follows:

Order	1	2	3	4	...	n
Number of Elements	1	$n+0_{\mathbb{Z}_{n,2}}$	$0_{\mathbb{Z}_{n,3}}$	$0_{\mathbb{Z}_{n,4}}$	...	$0_{\mathbb{Z}_{n,n}}$

Let G be a group with the above order structure.

Let $a \in G$ be of order n and b an element which is not a power of a. Then b and ba must be of order 2.

Therefore $baba = 1$, and $ba = a^{-1}b$ follows.

Having $a^n = b^2 = 1$ and $ba = a^{-1}b$, the 2n-element group G must be isomorphic to D_n. ∎

Proposition 54. *For all* n,

$$\text{i)} \quad M_n < N_{n+1} < N_n$$

$$\text{ii)} \quad N_{n+1}M < N_n M$$

$$\text{iii)} \quad B_{2^{n+1}m} N_{n+1} < B_{2^{n+1}m} N_n$$

$$\text{iv)} \quad B_{2^{n+1}m} N_{n+1} M < B_{2^{n+1}m} N_n M$$

If n *is even but not a power of* 2, *then*

$$\text{v)} \quad B_n M_m < B_n N_{m+1}$$

$$\text{vi)} \quad B_n M < B_n N_m M$$

Proof. To see the first part of i), use any D_p with p an odd prime. For example $D_3 = S_3$, the group of all permutations of three objects. Since

N_{n+1} is a subvariety of M_n, $M_n \leq N_{n+1}$ is obvious. Moreover, the SIN group D_3 is in M_n but not in N_{n+1}. Proposition 9 gives the result. Since D_{2n+1} is also SIN in Gru and $D_{2n+1} \in N_{n+1}M \setminus N_n$ one obtains the rest of i) and ii), iii) and iv). Suppose now that n is even and not a power of two. There is a prime $p \neq 2$ such that $2p|n$. Then the SIN group D_p is in $B_nM \subseteq B_nM_m$ but neither in B_nN_{m+1} or in B_nN_mM as D_p is not nilpotent of any length. ∎

Corollary.

i) Gru < N_m iii) $B_n < B_nN_m$ v) $N_2 <$ Ab $(= N_1)$

ii) $M < N_mM$ iv) $B_nM < M_nN_2$ vi) $B_{4m}N_2 < Ab_{4m} (= B_{4m}N_1)$

For even n *which is not a power of* 2.

Also $N_3M < N_2$, $B_8M < B_8N_2$, $B_{8n}N_3 < B_{8n}N_2$, $B_{16n}N_4 < B_{16n}N_3 < B_{16n}N_2 < Ab_{16n}$. ∎

Corollary. *For all even* n, $n \neq 2$ *we have:*

$$B_n < Ab_n. \blacksquare$$

For our next results we use Propositions 2 and 3 of [28] which in fact show that the groups S_4 and A_5 (respectively the group of all permutations of four objects and the group of all *even* permutations of five objects) are SIN in Gru.

Proposition 55.

i) $M_2 < M$, ii) $B_{12n}M_2 < B_{12n}M$, iii) $B_{12}M_2 < B_6M_2$.

Proof. The group S_4 is 2-metabelian but not metabelian, and it is of exponent 12 but not of exponent 6.

We know by [28] that S_4 is a SIN group and our result follows from Proposition 9.

To see that S_4 is not metabelian:

Call $a = (1,2,3,4)$, $c = (1,3,2)$, $b = c^{-1}a^2c = (1,4)(2,3)$. Then $[a,b][c,a] = (2,3,4) \neq (1,2,4) = [c,a][a,b]$.

To see that S_4 is 2-metabelian:

S_4 is presented by $a^4 = b^2 = c^3 = 1$, $bab = a^{-1}$, $c^{-1}a^2c = b$, $c^{-1}bc = a^2b$, $a^{-1}ca = c^2a^2b$.

Call $T := [S_4, S_4]$ the derived group.

By adding the equation "$xy \approx yx$" we would obtain $S_4/T \cong \mathbb{Z}_2$ as $a = a^{-1}$ means $a^2 = 1 = b = c$.

Therefore T has 12 elements, and must have a normal subgroup S of order 3 or 4. T/S is then abelian and therefore $[T:T] \subset S$ is also abelian. ∎

Proposition 56.

$$\text{i)} \quad \text{Gru} < M_n$$

$$\text{ii)} \quad B_{30m} < B_{30m}M_n$$

Proof. The group A_5 is a SIN group according to [28]. It is simple and therefore not n-metabelian for any n. Its exponent is clearly 30. ∎

By now one could be tempted to think that a corollary of Proposition 51 will be: "If $\mathcal{V} \subset \mathcal{W} \subset$ Gru then $\mathcal{W} < \mathcal{V}$." However, the order structure does not characterize the group and therefore one cannot conclude that all groups are SIN.

For example, one sees in [40] the following groups of order 16: $\mathbb{Z}_4 \times \mathbb{Z}_2 \times \mathbb{Z}_2$, $16\Gamma_2 b$ and $16\Gamma_2 c$, with the common order structure

1	2	4
1	7	8

, and $\mathbb{Z}_4 \times \mathbb{Z}_4$, $16\Gamma_2 a$ and $16\Gamma_2 c_2$ with

1	2	4
1	3	12

.

We do not know if these groups are SIN groups or not. It will have to be decided without the help of Proposition 51.

We, however, know of many groups which are *not* SIN groups, because our next proposition will show infinitely many examples of two different varieties which are equal in L.

A Lemma is first needed:

Lemma. *Let* $\mathcal{U}, \mathcal{V}$ *be varieties of groups such that* $\mathcal{U} \subset N_2$ *and* $\mathcal{V} \subset \mathrm{Ab}$. *If the term* $x^n y^m [x,y]^p$ *of* $F_2\mathcal{U}$ *defines a* $\mathcal{V}$*-operation, then* $x^n = x$, $y^m = y$ *and* $[x,y]^{2p+1} = 1$. *Moreover the term* $xy[x,y]^p$ *defines for all* p, *an associative operation.*

Proof. Consider $a*b = a^n b^m [a,b]^p$. Since $*$ is an abelian group operation $x \approx x*1 \approx x^n 1^m 1^p \approx x^n$ and similarly $y = y^m$. Since $*$ is also a commutative operation

$$xy[x,y]^p = x*y = y*x = yx[y,x]^p = xy[y,x]^{p+1}$$

and $[x,y]^{2p+1} \approx 1$ follows.

To see that $*$ is associative, one first checks that for a 2-nilpotent group:

$$[xy[x,y]^p,z] = [xy,z] \quad \text{and} \quad [x,yz[y,z]^p] = [x,yz].$$

Then

$$(x*y)*z = xy[x,y]^p z[xy[x,y]^p,z]^p = xyz[x,y]^p[xy,z]^p$$

and

$$x*(y*z) = xyz[y,z]^p[x,yz\,[y,z]^p]^p = xyz[y,z]^p[x,yz]^p.$$

Moreover

$$[x,y][xy,z] = x^{-1}y^{-1}xy\cdot y^{-1}x^{-1}z^{-1}xyz = x^{-1}y^{-1}z^{-1}xyz =$$
$$x^{-1}[y,z]z^{-1}y^{-1}xyz = [y,z]x^{-1}z^{-1}y^{-1}xyz = [y,z][x,yz]. \blacksquare$$

Proposition 57.

$$B_nN_2 =_L Ab_n \quad \textit{if and only if} \quad n = 2 \quad \textit{or} \quad n \quad \textit{is odd.}$$

Proof. For $n = 2$ this is obvious since $B_2 = Ab_2$. So let n be an odd number. Since Ab_n is a subvariety of B_nN_2, $B_nN_2 \le Ab_n$ is clear. By the lemma the term $xy[x,y]^p$ for $n = 2p+1$ defines a derived abelian group structure $a*b = [a,b]^p$ and since $a^{(*)k} = a^k$ it is also of exponent n. Conversely suppose n is an even number different from 2. If we had $Ab_n \le B_nN_2$ by the lemma, there would be a term $xy[x,y]^p$ of $F_2(B_nN_2)$ which would define an Ab_n-derived operation and for which $[x,y]^{2p+1} \approx 1$. Since the order of $[x,y]$ is n, $n|2p+1$ which is a contradiction. $\blacksquare$

J.A. de la Peña obtained the above result independently when he tried to generalize our earlier version of this Proposition. See [66]. We showed at that time only $B_3N_2 =_L Ab_3$.

$F_2(B_nN_2)$ has n^3 elements of the form $x^my^p[x,y]^q$ where $0 \le m,p,q < n$. The multiplication rule is

$$x^my^p[x,y]^q \cdot x^{m'}y^{p'}[x,y]^{q'} = x^{m+m'}y^{p+p'}[x,y]^{p+p'-mn'}.$$

For an element $x^my^p[x,y]^q$ we have

$$x^my^p[x,y]^q \cdot x^{n-m}y^{n-p}[x,y]^{n-q-mp} = x^ny^n[x,y]^{q+n-q-mp+mp}$$
$$= x^ny^n[x,y]^n = 1.$$

Therefore this set of elements with the above multiplication is a group and $\mathrm{ord}[x,y] = n$.

(A finite cancellative semigroup is a group.)

Note that we have shown $B_nN_2 < Ab_n$ for all even n, $n \ne 2$.

From the proof of Proposition 57 one sees that if P_n is the variety of groups defined by $[x,y]^{2n+1} \approx 1$, $P_nN_2 =_L Ab$ follows for all n.

As a further consequence of the lemma, a new proof for $N_2 < Ab$ is obtained, since $[x,y]$ is torsion-free in $F_2(N_2)$.

From Proposition 57 we again obtain that $B_n < Ab_n$ for all even n, $n \neq 2$.

Our next goal is to find out for which values of n, $B_nN_3M = Ab_n$. We also hope to find odd values of n such that $B_n < Ab_n$. For this purpose we shall use the well known fact that the "commutator collecting process" allows a unique representation of each element of $F_m(MN_3)$ and of $F_m(B_nN_3M)$ when n is prime, $n > 3$, as a product of powers of the m generators, $\binom{m}{2}$ commutators of them of length 2 and $2\binom{m+1}{3}$ commutators of length 3. See for example [58].

Proposition 58.

$$B_3 < Ab_3.$$

Proof. We shall show that no term of F_2B_3 (the free group of rank 2 in the variety B_3) defines an Ab_3-structure. Notice that the group F_2B_3 is 2-nilpotent.

As in our lemma before Proposition 56, we see that only terms $xy[x,y]^p$ can define an operation with identity, and that for commutativity we need $2p + 1 = 3$. Therefore $p = 1$ and the only term we have to analyze is $xy[x,y]$.

Now we know from [52] that B_3 is a metabelian variety and $[x,y,z] \approx [z,x,y] \approx [y,z,x]$ but $[x,y,z] \approx 1$ is not a law.

Let us consider the associativity:

$$(x*y)*z = xy[x,y]z[xy[x,y],z] = xyz[x,y][x,y,z][xy,z][x,y,z]$$
$$= xyz[x,y][x,y,z][x,z][y,z][x,z,y][x,y,z]$$
$$= xyz[x,y][x,z][y,z][x,y,z]$$

and

$$x*(y*z) = xyz[x,y][x,z][y,z].$$

Since $[x,y,z] \approx 1$ is not a law, $*$ is not associative. Therefore $Ab_3 \not\leq B_3$ and the result follows. ∎

In order to obtain a proposition similar to Proposition 56 we need a concrete functor $F : MN_3 \to Ab$ or $F : B_nN_3M \to Ab_n$. Therefore we look for an appropriate term first in MN_3 which should define an Ab-operation. Our discussion here is similar to that in the proof of the lemma preceding Proposition 57.

Considering once more the "commutator collecting process," we just look at terms of the form:

$$x^ny^m[x,y]^p[x,y,x]^q[x,y,y]^r$$

Suppose this term defines an abelian group operation $*$. Since $1*1 = 1$, 1 is the identity with respect to $*$.

One has $x = x*1 = x^n$ and $y = 1*y = y^m$, and since $*$ is also commutative:

$$xy[x,y]^p[x,y,x]^q[x,y,y]^r = x*y = y*x$$
$$= yx[y,x]^p[y,x,y]^q[y,x,x]^r$$
$$= xy[y,x]^{p+1}[y,x,y]^q[y,x,x]^r .$$

And therefore:

$$[x,y]^{2p+1}[x,y,x]^{q+r}[x,y,y]^{q+r} = 1.$$

This means that given a (necessarily) odd number n, if we can find p,q and r such that

$$n = 2p + 1 = q + r,$$

the term

$$xy[x,y]^p[x,y,x]^q[x,y,y]^r$$

will define a commutative operation in B_nN_3M with an identity 1. (The operation is also of exponent n since $x^{(*)k} = x^k$ is clear.) This will of course be also achieved if just $n|2p+1$ and $n|q+r$.

Our problem is now reduced to finding for which of the values of p,q and r satisfying the above condition, $*$ also defines an associative operation. For this purpose, the following commutator identities which apply in MN_3 are useful: (Recall that in our notation $[x,y] = x^{-1}y^{-1}xy$.)

(1) $yx = xy[y,x]$

(2) $[xy,z] = [x,z][y,z][x,z,y]$

(3) $[x,y,z] = [y,x,z]^{-1}$

(4) $[x,y,z][y,z,x][z,x,y] = 1$

(5) $[x,y]^pz = z[x,y]^p[x,y,z]^p$

(6) $[x[y,z,w],v] = [x,v]$

(7) $[xy[x,y]^p,z] = [x,z][y,z][x,z,y][x,y,z]^p$

Since $(x*y)*z$ and $x*(y*z)$ will be words of $F_3(MN_3)$ according to [58], we choose to represent them with the following basic commutators of length 3:

$$[x,y,y],\quad [x,y,x],\quad [x,y,z]$$
$$[z,y,y],\quad [z,y,x],\quad [z,y,z]$$
$$[z,x,x],\quad [z,x,z]$$

By repeated applications of (1) through (7) one obtains:

$$(x*y)*z = xyz[x,y]^p[x,z]^p[z,y]^{-p}[x,y,y]^r[x,y,x]^q[x,y,z]^{p^2+2p+q}$$
$$[z,y,y]^{-q}[z,y,x]^{-(p+2q)}[z,y,z]^{-r}[z,x,x]^{-q}[z,x,z]^{-r},$$

and

$$x*(y*z) = xyz[x,y]^p[x,z]^p[z,y]^{-p}[x,y,y]^r[x,y,x]^q[x,y,z]^{p+2r}$$
$$[z,y,y]^{-q}[z,y,x]^{p^2-r}[z,y,z]^{-r}[z,x,x]^{-q}[z,x,z]^{-r}.$$

The condition for associativity is therefore:

$$[x,y,z]^{p^2+p+q-2r} = [z,y,x]^{p^2+p+2q-r}.$$

We shall soon prove a stronger result than the next one, but we include it here as a lemma since its proof is a similar version of the next proofs and the result is used in them.

Lemma. *If* n *is a prime number, and* $n > 3$, *then*

$$B_nN_3N = {}_LAb_n.$$

Proof. Similar to Proposition 57. Since $B_nN_3M \leq Ab_n$, we seek a term $xy[x,y]^p[x,y,x]^q[x,y,y]^r$ which defines an abelian group operation.

By [58] we know that the commutators $[x,y]$, $[x,y,x]$ and $[x,y,y]$ and also $[x,y,z]$ and $[z,y,x]$ are independent. In other words, $[x,y]^{e_1}[x,y,x]^{e_2}[x,y,y]^{e_3} = 1$ imply $e_1 = e_2 = e_3 = 0$ modulo n, and $[x,y,z]^{e_4}[z,y,x]^{e_5} = 1$ imply $e_4 = e_5 = 0$ modulo n.

If we define $x*y = xy[x,y]^p[x,y,x]^q[x,y,y]^r$ we can assume that $0 \leq p,q,r < n$. Since we look for $*$ to be commutative, the commutativity condition $n|2p+1$ and $n|q+r$, yields $2p < 2n$, $2p+1 \leq 2n$, and since $2n$ is even, $2p+1 < 2n$. Therefore $n = 2p+1$. Similarly $q+r < 2n$ and $n|q+r$ yields $n = q+r$ or $q+r = 0$.

The condition for associativity is

$$n|p^2 + p + q - 2r$$
$$n|p^2 + p + 2q - r.$$

Since $r = n-q$ (as $q+r = 0$ leads to a non-associative operation), one obtains $q-2r = 3q-2n$ and $2q-r = 3q-n$, and therefore the associativity conditions collapse into

$$n|p(p+1) + 3q.$$

We now show that q can be found:

Let k,ℓ be integers such that $kn-3\ell = 1$. Then

$$p(p+1) = 1\cdot p(p+1) = kp(p+1)n - 3\ell p(p+1)$$

and

$$kp(p+1)n = p(p+1) + 3\ell p(p+1).$$

Let $0 < q < n$ be such that $\ell p(p+1) \equiv q$ modulo n, and $\ell p(p+1) - q = mn$. Then

$$n|(kp(p+1)-3m)n = p(p+1) + 3q.$$

Define now $r := n-q$, and we have found the term which we were looking for. Therefore $Ab_n \leq B_nN_3M$ and the lemma follows. ∎

Lemma. *Let* n *be an odd positive integer and* p *a prime number such that* $p|n$. *Then* p *divides the order of* $[x,y,z]$ *in* $F_3(B_nN_3M)$.

Proof. By [58] we know that the order of $[x,y,z]$ in $F_3(B_pN_3M)$ is p, and $F_2(B_pN_3M)$ is a quotient of $F_3(B_nN_3M)$. Therefore $p|\text{order}([x,y,z])$. ∎

Before we state our next proposition, recall that $B_3 = B_3N_3M$.

Proposition 59. *Let* n *be an odd positive integer. Then*

$$B_nN_3M =_L Ab_n \qquad \textit{if and only if } 3\nmid n.$$

Proof. Suppose $3\nmid n$, and let π be the order of $[x,y,z]$. Therefore $3\nmid\pi$, and we proceed as in the lemma for prime numbers. We can assume $0 \leq p,q,r < \pi$ and the commutativity condition leads to $\pi = 2p+1$ and $\pi = q+r$ or $q+r = 0$.

The associativity condition here is $\pi|p(p+1)+3q$ and since only $(n,3) = 1$ was needed in the proof of the lemma, $(\pi,3) = 1$ here allows us to find q and therefore also $r = \pi-q$.

Conversely, suppose $3|n$. By the lemma, $3|\pi$. If a term $xy[x,y]^p[x,y,x]^q[x,y,y]^r$ defines an abelian group operation $*$, the commutativity condition requires $3|2p+1$ (as $3|n$ and $n|2p+1$ yields $3|2p+1$).

Therefore $p = 3t+1$ for some integer t. But then $3\nmid p(p+1) + 3q = (3t+1)(3t+2) + 3q$ for any q. This contradicts the associativity of $*$. ∎

Proposition 60. *If* n *is a positive integer and* $2|n$ *or* $3|n$, *then*

$$B_n < Ab_n.$$

Proof. For n even, we use the corollary to Proposition 54. For $3|n$, Proposition 59 yields $B_n \leq B_nN_3M < Ab_n$. ∎

Note that if n is odd and $3\nmid n$ we know that $B_nN_3M =_L Ab_n$, but it is still an open problem whether $B_n < B_nN_3M$, or rather for which values of n this happens.

Accordingly, we have

Problem 1.

i) *For which odd values of* n *with* $3\nmid n$ *is* $B_n < Ab_n$ *?*

ii) *For which primes* p, $p \neq 3$, *is* $B_p < Ab_p$ *?*

We believe that a comparison of B_nN_mM with Ab_n may yield information, when m is sufficiently large. However, a deep understanding of the behavior of the commutators of great length will be required to be able to handle the otherwise massive computations involved, to obtain the commutativity and associativity conditions.

A combination of Propositions 57 and 59 gives a complete solution to the problem: *For which values of* n *is* $B_nN_3M < Ab_n$? Indeed:

Proposition 61.

$$B_nN_3M < Ab_n \quad \textit{precisely if} \quad n \neq 2 \ \textit{even or} \ 3|n.$$

Proof. For $n \neq 2$, even: $B_nN_3M \leq B_nN_2 < Ab_n$. ∎

Let us return to the even values of n to find information about when $B_nN_3M < B_nN_2$.

We look first for a $F_2(MN_3)$ term which could define a N_2-group operation. Only terms

$$xy[x,y]^p[x,y,x]^q[x,y,y]^r$$

for which $[x,y,z]^{p(p+1)+q-2r} = [z,y,x]^{p(p+1)+2q-r}$ have to be considered, as * still needs to have an identity and be associative.

To obtain the condition for 2-nilpotency, using the following commutator identities,

(1') $[x^{-1},y] = [y,x][y,x,x^{-1}]$

(2') $[x,y^{-1}] = [y,x,[y,x,y^{-1}]]$

(3') $[x^{-1},y^{-1}] = [x,y][x,y,x^{-1}][x,y,y^{-1}]$

(4') $[x,y,x^{-1}] = [x,y,x]^{-1}$

(5') $[x,y,y^{-1}] = [x,y,y]^{-1}$

(6') $[x^{-1},y^{-1},x^{-1}] = [x,y,x^{-1}]$

(7') $[x^{-1},y^{-1},y^{-1}] = [x,y,y]^{-1}$

(8') $[x,x^{-1}y^{-1}[x,y]^p] = [y,x][x,y,y][x,y,x]^{-p}$

(9') $[y,x^{-1}y^{-1}[x,y]^p] = [x,y][x,y,x]^{-1}[x,y,y]^{-(p+1)}$

(10') $[x,y,x^{-1}y^{-1}[x,y]^p] = [x,y,x]^{-1}[x,y,y]^{-1}$

(11') $[xy[x,y]^p,x^{-1}y^{-1}[x,y]^p] = [x,y,x]^{-(2p+1)}[x,y,y]^{-(2p+1)}$

one obtains:

$$[x,y]_* = [x,y]^{2p+1}[x,y,x]^{p(2p+1)}[x,y,y]^{p(2p+1)}$$

and

$$[x,y]_* * z = z[x,y]^{2p+1}[x,y,x]^{p(2p+1)}[x,y,y]^{p(2p+1)} \cdot [[x,y]^{2p+1},z]^{p+1}$$

whereas

$$z*[x,y]_* = z[x,y]^{2p+1}[x,y,x]^{p(2p+1)}[x,y,y]^{p(2p+1)} \cdot [[x,y]^{2p+1},z]^{-p}.$$

Therefore the condition for 2-nilpotency is

$$[x,y,z]^{(2p+1)^2} \approx 1.$$

This has given an alternative proof to the following result which is part of Proposition 54 i) and ii):

$$N_3 \le MN_3 \le N_2 .$$

Proof. Since by [58], the commutator $[x,y,z]$ is torsion-free, the condition $[x,y,z]^{(2p+1)^2} \approx 1$ is not satisfied for any p. ∎

Lemma. *Let* p *be a prime number and* $p > 3$. *If* $2p|n$, *then the order of* $[x,y,z]$ *in* $F_3(B_nN_3M)$ *is even.*

Proof. Since $F_3(B_{2p}N_3M) \in B_nN_3M$ it is sufficient to show that the order π of $[x,y,z]$ in $F_3(B_{2p}N_3M)$ is even. We know from [58] that the order of $[x,y,z]$ in $F_3(B_pN_3M)$ is p. Therefore, if we assume that π is odd, $\pi = p$. But then we can proceed as in the proof of Proposition 58 (which depends on the odd order of $[x,y,z]$ and its indivisibility by 3) and show that $B_{2p}N_3M = {}_LAb_{2p}$. But this contradicts Proposition 61, according to which $B_{2p}N_3M \le B_{2p}N_2 < Ab_{2p}$. ∎

Proposition 62.

$$B_nN_3M < B_nN_2 \quad \textit{if and only if} \quad n \neq 2 \ \textit{is even or} \ 3|n.$$

Proof. For odd n the proof is like in Proposition 59 since the 2-nilpotency condition $3|(2p+1)^2$ leads to $3|2p+1$ and one can continue verbatim. For the converse in the case n odd, Proposition 59 proves even more.

So let n be an even number and $n \neq 2$. If $4|n$, $[x,y,z] \neq 1$ as it is so already in B_4N_3M, see [27], therefore $2|\mathrm{ord}(x,y,z)$ in $F_3(B_nN_3M)$ and the condition $[x,y,z]^{(2p+1)^2} \approx 1$ cannot be satisfied for any p.

Likewise, if there is a prime $q > 3$ such that $q|n$, from [58] it follows that $[x,y,z] \neq 1$, and from the lemma $2|\mathrm{ord}(x,y,z)$ in $F_3(B_nN_3M)$ leading to the same result.

Suppose then that $n = 2 \times 3^m$ for some $m \ge 1$. We shall show that no term of $F_2(B_{2\times 3^m}N_3M)$ which satisfy the 2-nilpotency condition, will also define an associative operation. From our lemma before Proposition 59, we know that 3 divides the order $[x,y,z]$ already in $F_3(B_{3^m}N_3M)$. So it also does in $F_3(B_{2\times 3^m}N_3M)$.

If the 2-nilpotency condition is satisfied for a term

$$xy[x,y]^p[y,x,x]^q[y,x,y]^r$$

one has

$$[x,y,z]^{(2p+1)^2} \approx 1$$

and therefore $3|2p+1$. One now continues as in Proposition 59, using the associativity condition found in the proof of the previous lemma for prime numbers. Indeed, without using the commutativity condition $n|q+r$, one still has associativity conditions

$$3|p(p+1) + q - 2r \quad \text{and} \quad 3|p(p+1) + 2q - r$$

Therefore

$$3|(2q-r) - (q-2r) = q + r$$

which means $q - 2r = 3q - 6h$ for some h, and the associativity condition is changed to

$$3|p(p+1)$$

which together with $3|2p+1$ provides the contradiction pointed out in Proposition 59. ∎

In our Proposition 54 we solved all but two cases of the following.

<u>Problem 2</u>.

For which values of n is $B_nM < B_nN_3M$?

After our contribution to this problem, the still open question looks as follows:

i) $B_4M < B_4N_3M$?

ii) $B_8M < B_8N_3M$?

Using the information contained in [40] may help to locate adequate SIN groups if the answer is "yes."

We know from [40] that there are only 2 groups of order $\leq$ 64 and exponent 4 which are metabelian but <u>not</u> 3-nilpotent. Those are $G_1 = 64\Gamma_{23}a_1$ and $G_2 = 64\Gamma_{23}a_2$. Unfortunately we do not know if these groups are SIN in Gru, since the following groups have the same order structure as G_1 : $64\Gamma_{10}b_3$, $64\Gamma_{10}b_4$, $64\Gamma_{10}c_1$, $64\Gamma_{10}c_2$, $64\Gamma_{10}c_3$, $64\Gamma_{10}c_4$, $64\Gamma_{13}a_3$ and $64\Gamma_{25}a_2$, and the order structure

2	4
11	52

of G_2 is also shared by: $64\Gamma_{10}a_2$, $64\Gamma_{10}b_5$, $64\Gamma_{10}c_5$, $64\Gamma_{10}c_6$ and $64\Gamma_{13}a_4$.

There may be, of course, an argument why G_1 or G_2 could be a SIN group, but pursuing it would lead us outside our present work. On the other hand, one could look for terms in $F_2(B_4M)$ which would give a B_4N_3M operation. For this purpose there is much information about groups of exponent 4 and 8 obtained by the people working on the Burnside Problem, [10] and [27], and it may be useful. We consider this research the subject of a separate effort.

We finish our discussion on the location of variation of groups in L with a list of open problems.

<u>Problem 3</u>. *For which* m *and* n *is* $B_m < B_m N_n$?

Our contribution: Proposition 54 for even m which is not a power of 2, and all n. If m is a power of 2, say 2^k for $n < k$,
Is $B_{2^k} < B_{2^k} N_n$ for some $n \geq k$?

<u>Problem 4</u>. *For which* n *is* $B_n < B_n N_2$?

Our contribution: If n is even, see Problem 3.
If $n = 3$ $B_3 < Ab_3 =_L B_3 N_2$, use Propositions 57 and 58.
If n is odd $n \neq 3$ and $3|n$

$B_n \leq B_n N_3 M < Ab_n =_L B_n N_2$ use Propositions 57 and 59.

<u>Problem 5</u>. *For which* n *is* $B_n M < B_n N_2$?

Our contribution: When n is odd: For $3|n$.
If $n \neq 2$ even, $B_n M \leq B_n N_3 M < B_n N_2$ Proposition 62.
If n odd, $n \neq 3$, $B_n M \leq B_n N_3 M < Ab_n =_L B_n N_2$ Propositions 57 and 59.
If $n = 3$, $B_3 = B_3 M < Ab_3 =_L B_3 N_2$ Propositions 57 and 58.

<u>Problem 6</u>. *For which* m *and* n *is* $B_m M_n < B_m N_{n+1}$?

Our contribution: Proposition 54.

<u>Problem 7</u>. *For which* n *is* $B_n < B_n M$?

Our contribution:
If $12|n$ $B_n \leq B_n M_2 < B_n M$ Proposition 55.
If $30|n$ $B_n < B_n M$ Proposition 56.

<u>Problem 8</u>. *For which* n *and* m *in* $B_n < B_n M_m$?

Our contribution:
If $30|n$ Proposition 56.

<u>Problem 9</u>. $N_3 < N_3 M$?

For which n (if any) is $N_n < N_n M$?

Problem 10. *For which* n *is* $B_nN_3 < B_nN_2$?

Our contributions: See Propositions 54, 57, 58, 59 and 62. See Problem 5.

Problem 11.

i) Is there a maximum n such that $M < N_n$?

ii) Can $M_n < N_{n+1}$ be improved ?

iii) For which n is $M_{n+1} < M_n$?

iv) Is $M_3 < M_2$?

Problem 12.

i) For which n is $B_nM < B_nN_mM$ (all m) ?

ii) For which n is $B_nM < B_nN_3M$?

iii) For which *odd* n is $B_nM < B_nN_3M$?

iv) Is $B_4M < B_4N_3M$?

v) Is $B_8M < B_8N_3M$?

Our contribution: Proposition 54. (i) is solved for even n not a power of 2 and partial answer for powers of 2.)

Problem 13. *Propositions involving products of varieties of groups:*

Example – $A <_L B \Rightarrow A.C <_L A.B$?

Regarding Problem 1, ii), namely "is $B_5 < Ab_5$?", we think that the results of [58] may help to decide it. However, massive computations to check and obtain conditions for associativity, metabelianity, and perhaps n-nilpotency may be required unless a more powerful technique is discovered.

Problem 14. *Which is the highest value of* m *such that* $B_nN_mM = {}_LAb_n$ *if* n *is odd and* $3|n$?

See Proposition 59.

A solution to Problem 14 may also yield a contribution to Problem 1.

We now consider the categories of lattices. Their location in L can be nicely described by the following lemma, which is an analog to our lemma for the variety of groups given just before Proposition 19.

Lemma. *There are only two concrete functors of* Lat *into itself. The identity and the "dual interpretation" in which the terms* $x \wedge y$ *and* $x \vee y$ *are interchanged.*

Proof. There are only 4 possible binary terms, since the free algebra of rank 2, F_2Lat, in lattices has only 4 elements: a, b, $a \wedge b$, $a \vee b$. The term defined by "a" induces the derived operation $x \cdot y \approx x$ which is non commutative, and the term "b" induces $x \cdot y \approx y$ which is also not commutative, and therefore neither of them define a lattice operation. We are left with only the words $a \wedge b$ and $a \vee b$. Neither of them can define both lattice operations, because in such case $(x \wedge y) \vee x \approx x$ or $(x \vee y) \wedge x \approx x$ would not be satisfied. The only possibilities are then as stated in the lemma. ∎

Proposition 63. *Let* V *and* W *be varieties of lattices. Then* $V < W$ *in* L *if and only if* W *is a proper subvariety of* V *or the dual of* V. *If* V^d *is the dual of* V, *then* $V =_L V^d$.

Proof. Given $V \leq W$, by the lemma, we must have the concrete functor $F : W \to V$ defined by $F(A;\wedge,\vee) = (A;\wedge,\vee)$ or $F(A;\wedge,\vee) = (A;\vee,\wedge)$. Therefore $W \subseteq V$ or $W \subseteq V^d$. If now $W \nsubseteq V$, then $V \subsetneq W$ and $V \nsubseteq W^d$, i.e. $V^d \nsubseteq W^{dd} = W$. The converse is also clear and the second statement is a by-product. ∎

Corollary. *The partially ordered set of selfdual varieties of lattices occurs (as a poset) unaltered in* L. ∎

Remark. But not as a sublattice! Any variety of lattices generated by its finite members is $\wedge$-prime, by the Corollary after Proposition 18. There are $2^{\aleph_0}$ such varieties which are selfdual.

10. THE BOTTOM PART OF L.

We begin the study of the interval [Sem,Ab] but do not discuss here the part [Gru,Ab] since this has been done in §9.

We adopt the notation of [12], page 24, and define *quasigroups* as algebras of type $\langle 2,2,2\rangle$ with operations denoted by $\cdot, \backslash, /$ and satisfying the following identities

$$Q1 : x\backslash(x\cdot y) \approx y, \qquad (x\cdot y)/y \approx x$$

$$Q2 : x\cdot(x\backslash y) \approx y, \qquad (x/y)\cdot y \approx x$$

By Qgr we shall understand the variety of quasigroups. By MQ1 the variety of models of Q1 and by SMQ1 the subvariety of MQ1 for which the first operation is associative.

Proposition 64.

$$0 =_L \text{Sem} < \text{MQ1} \leq \text{Qgr}$$

Proof. A semigroup can be defined on each set by $xy \approx x$ and therefore by Proposition 10 in §1, $0 =_L \text{Sem} =_L \text{Set}$ is the zero of L.

Since Qgr is a subvariety of MQ1, MQ1 $\leq$ Qgr, and the inequality $0 <$ MQ1 is obvious from Proposition 10 and its Corollary. ∎

Proposition 65.

i) $C \not\leq$ Qgr

ii) $C \not\leq$ SMQ1

iii) Qgr < Gru

iv) MQ1 < SMQ1

Proof. Consider the algebra $Q = (\{a,b,c\},\cdot,\cdot,\cdot)$ where $\cdot$ is defined as

$\cdot$	a	b	c
a	b	a	c
b	a	c	b
c	c	b	a

(By using Q1 and Q2 one obtains / and \ which coincide in this case with $\cdot$.)

Since the permutation (a,b,c) is an automorphism of Q, Aut(Q) has at least 3 elements. However, any 3-element algebra A in C, Gru or SMQ1 has an automorphism group isomorphic to $\mathbb{Z}_2$. In C the constant has to be fixed under automorphisms and any 3-element algebra in Gru or SMQ1 is isomorphic to $\mathbb{Z}_3$ (finite cancellative semigroups are groups). Therefore File(Q) $\nleq$ File A.

Given that Gru $\nleq$ Qgr, Qgr $\nleq$ Gru is an obvious interpretation and iii) follows. For ii) notice that SMQ1 has the model L used below in Proposition 67, for $\mathcal{U}$ $\nleq$ SMQ1. ∎

<u>Proposition 66</u>. *The variety* C *is incomparable with* Qgr *and with* SMQ1.

<u>Proof</u>. We see that SMQ1 $\nleq$ C because the five element model A of C has Aut(A) $\cong \mathcal{S}_4$, while the only model with five elements in SMQ1 is isomorphic to $\mathbb{Z}_5$. (We stress again that finite models in SMQ1 are cancellative semigroups with respect to · and therefore groups.) Obviously File($\mathbb{Z}_5$) $\nleq$ File(A).

We give a different argument for Qgr $\nleq$ C. (It also provides a second proof for SMQ1 $\nleq$ C).

Suppose Qgr $\leq$ C. By the definition of $\leq$ in L, for each Qgr-operation: ·, /, \ there exist terms t(x,y), d(x,y), p(x,y) in F_2C such that if (A,·) is any algebra in C, then (A,t,d,p) is an algebra in Qgr.

Take the algebra

	c
a	a
b	a

in C.

Suppose t,d,p have been given. t(x,y) may be one of

i) t(x,y) = a

ii) t(x,y) = x

iii) t(x,y) = y

(We write xty instead of t(x,y)).

Now since ({a,b};t,d,p) ε Qgr one has, for "xty = a"

$$a = (atb)db = adb = (btb)db = b$$

For "xty = x",

$$a = bp(bta) = bpb = bp(btb) = b$$

and for "xty = y"

$$a = (atb)db = bdb = (btb)db = b$$

A contradiction in each of the three possible cases. ∎

Recall from our list of notation for varieties that the variety $\mathcal{U}$ is defined by the equation xy $\approx$ yx.

Proposition 67. *The varieties* $\mathcal{U}$ *and* SMQ1 *are incomparable.*

Proof. To see that $\mathcal{U} \not\leq$ SMQ1 consider the following algebra L of SMQ1. On the underlying set of F_2Sem, i.e. the set of non-empty words on two generators a,b, we define for the words v,w:

$$v/w = vw \text{ (juxtaposition) unless } v = \gamma w$$
$$(\gamma w)/w = \gamma$$
$$v \backslash w = vw \quad \text{unless} \quad w = v\delta$$
$$v \backslash (v\delta) = \delta$$

L has an involution h such that h(a) = b and h(b) = a, with no fixed points. It is now immediate that $\mathcal{U} \not\leq$ SMQ1, from P. Gumm's Theorem [37]. See our discussion on page 57, prior to Proposition 31.

SMQ1 $\not\leq \mathcal{U}$ follows from Proposition 66, since $\mathcal{U} \leq$ C and SMQ1 $\not\leq$ C. ∎

Corollary.

$$\text{SMQ1} < \text{Gru.} \blacksquare$$

The following two varieties of type $\langle 3 \rangle$ are important in our lattice. Mt_0 satisfying the equations:

$$m(x,x,y) \approx m(x,y,x) \approx m(y,x,x) \approx x$$

and

Mt_1 satisfying the equations:

$$p(x,x,y) \approx p(y,x,x) \approx y.$$

As is well known, the equations defining Mt_1 are the Mal'tsev conditions for permutability of congruences.

Because of its symmetry to Mt_0, we also consider Mt_2 of the same type but satisfying:

$$p(x,x,y) \approx p(x,y,x) \approx p(y,x,x) \approx y$$

Proposition 68. *Let* M *be one of the* Mt_i, i = 0,2. *Let* $\mathcal{V}$ *be a variety of type* $\langle 2 \rangle$ *and* $0 < \mathcal{V}$. *Then* $\mathcal{V} \not\leq$ M.

Proof. Suppose there is a concrete functor F : M → $\mathcal{V}$. Then a binary M-term of F_2M will interpret the binary operation of $\mathcal{V}$. But F_2M = {a,b} and no projection can interpret the binary operation of $\mathcal{V} > 0$. ∎

Proposition 69.

$$C \not\leq Mt_1.$$

Proof. We consider the C-operation unary. F_1M = {a} and f(x) = x is not a constant operation. ∎

Proposition 70.

i) $Mt_0 < Lat$

ii) $Mt_1 < Qgr$

iii) $Mt_0 \,||\, Mt_1$

Proof. For i) we first obtain $0 < \mathcal{U} \leq Lat$, where a forgetful functor forgetting one of the lattice operations provides $\mathcal{U} \leq Lat$, and $0 < \mathcal{U}$ is obvious. An application of Proposition 68 yields the desired result. For ii) we obtain $\leq$ from the polynomial $p(x,y,z) = (x/(y\backslash y))(y\backslash z)$ given in [45] page 366. And to obtain $<$ we give a 3-element model of Mt_1 with 3 different two-element submodels. (Such situations do not happen in Qgr and we apply Proposition 9.)

On the set $\{a,b,c\}$ we define the ternary operation $p(x,y,z)$ as follows:

$p(a,a,a) = a$	$p(b,a,a) = b$	$p(c,a,a) = c$
$p(a,a,b) = b$	$p(b,a,b) = b$	$p(c,a,b) = c$
$p(a,a,c) = c$	$p(b,a,c) = b$	$p(c,a,c) = c$
$p(a,b,a) = a$	$p(b,b,a) = a$	$p(c,b,a) = c$
$p(a,b,b) = a$	$p(b,b,b) = b$	$p(c,b,b) = c$
$p(a,b,c) = a$	$p(b,b,c) = c$	$p(c,b,c) = c$
$p(a,c,a) = a$	$p(b,c,a) = b$	$p(c,c,a) = a$
$p(a,c,b) = a$	$p(b,c,b) = b$	$p(c,c,b) = b$
$p(a,c,c) = a$	$p(b,c,c) = b$	$p(c,c,c) = c$

The lattice of subalgebras of $M = (\{a,b,c\},p)$ coincides with the lattice of subsets of its underlying set. For iii) we notice that the group $\mathbb{Z}_2 \times \mathbb{Z}_2$ has a non-distributive congruence lattice  and therefore $Mt_0 \not\leq Mt_1$. And the chain $a < b < c$ as a lattice has two congruences $\theta_1 = \{(a,a),(b,b),(c,c),(a,b),(b,a)\}$ and $\theta_2 = \{(a,a),(b,b),(c,c),(b,c),(c,b)\}$ such that $(a,c) \in \theta_1 \circ \theta_2$ and $(a,c) \notin \theta_2 \circ \theta_1$. Therefore θ_1 and θ_2 do not permute and $Mt_1 \not\leq Mt_0$ follows. ∎

It is obvious that $Mt_3 \leq Mt_0$ and $Mt_1 \leq Mt_2$ as Mt_0 is a subvariety of Mt_3 and Mt_2 of Mt_1. On the other hand, since $\mathbb{Z}_2 \times \mathbb{Z}_2$ is a model of $\mathcal{U}$, $Mt_0 \not\leq \mathcal{U}$ and we get the following

Corollary.

$$\mathcal{U} < Lat. \blacksquare$$

The variety W in our notation is of type $\langle 2 \rangle$ and its defining equations are $xy \approx yx$ and $(xy)(uv) \approx (xu)(yv)$.

Proposition 71. $Sem < \mathcal{U} < W$

Proof. As we just mentioned in the proof of the previous Proposition, Sem = 0 < $\mathcal{U}$ is obvious. (Use the corollary to Proposition 9 or just the algebra

	a	b
a	a	b
b	b	b

which is SIN in Sem.) For $\mathcal{U}$ < W we consider the algebra L defined on {a,b,c} by

∘	a	b	c
a	a	a	c
b	a	b	b
c	c	b	c

This algebra belongs to $\mathcal{U}$ but not to W, as

$$(a \circ c) \circ (b \circ c) = c \circ b = b \neq c = a \circ c = (a \circ b) \circ (c \circ c).$$

The permutation (a,b,c) is an automorphism of L of order 3 and has 3 different one-element subalgebras plus another 3 two-element subalgebras.

We claim that for every 3-element model A of W, File(L) $\not\leq$ File(A).

To show this, we would be interested in those 3-element algebras of W obtained by filling the table

	a	b	c
a	a		
b	·	b	
c	·	·	c

commutatively with different elements in the three empty places, as is required for ϕ = (a,b,c) to be an automorphism.

There are then 6 possibilities. We denote by "xyz" the algebra described by the following operation table:

	a	b	c
a	a	x	y
b	x	b	z
c	y	z	c

where x,y,z ε {a,b,c}.

The algebras now denoted by "abc," "cab" and "bca" have an identity and therefore no automorphism of order 3. The algebras "acb" and "bac" are isomorphic to L, and therefore do not belong to W. The algebra "cba" has no two-element subalgebras. ∎

The following general result will be useful to prove several of the subsequent propositions. We mean by C the variety of constant semigroups.

Proposition 72. *Let* $\mathcal{A}$ *and* $\mathcal{B}$ *be varieties. Let* $\mathcal{A}$ *be a selfdual variety of semigroups or monoids and let* $\mathcal{B}$ *have a reduct* $\mathcal{B}_r \subset \mathcal{A}$ *such that* $C \neq \mathcal{B}_r > 0$. *If* $(\mathbb{Z},+) \varepsilon \mathcal{A}$, *then* $\mathcal{A} < \mathcal{B}$.

Proof. Since $\mathcal{A} \leq \mathcal{B}$ is obvious, suppose $\mathcal{B} \leq \mathcal{A}$. There is a term $\alpha(x,y)$ in $F_2\mathcal{A}$ which defines a $\mathcal{B}$-operation. Let $\alpha(x,y) = x^{n_1}y^{m_1}x^{n_2}\ldots$ be one of its representations. On $(\mathbb{Z},+) \varepsilon \mathcal{A}$ this term defines the operation $a*b = Na + Mb$, where $M = \sum m_i$ and $N = \sum n_i$.

The associativity of * yields

$$M = (0*0)*1 = 0*(0*1) = M^2$$

$$N = 1*(0*0) = (1*0)*0 = N^2$$

and so $M = 0$ or 1, $N = 0$ or 1 and the only possibilities are:

$$\alpha(x,y) = x, \quad \alpha(x,y) = y, \quad \alpha(x,y) = xy \quad \text{and} \quad \alpha(x,y) = yx.$$

($M = N = 0$ would lead to a constant operation for $\mathcal{B}$.)

Since $\mathcal{B}_r > 0$, the first two are impossible, and if yx is an interpretation of $\mathcal{B}$ in A, then so is xy because we may combine yx with the interpretation yx of A in A. Therefore, it will be enough to show that xy does not interpret $\mathcal{B}$ in A, i.e. that $A \not\subseteq \mathcal{B}_r$. But this is given by our hypothesis $\mathcal{B}_r \subset A$. ∎

Our next proposition will refer to the following varieties of semigroups. V_n defined by the equation $yx_1\ldots x_n z \approx zx_1\ldots x_n y$, Semc, commutative semigroups, and Semℓ, the variety of semilattices.

Let $F_\omega V_n$ denote the V_n-free algebra of rank ω, and S_n the group of permutations of n objects.

<u>Lemma</u>. *If* $a_1 \cdot a_2 \cdot \ldots \cdot a_{n+3} \in F_\omega V_n$ *and* $\sigma \in S_{n+3}$, *then* $a_1 \cdot a_2 \cdot \ldots \cdot a_{n+3} = a_{\sigma(1)} \cdot a_{\sigma(2)} \cdot \ldots \cdot a_{\sigma(n+3)}$.

<u>Proof</u>. It is enough to show it for

$$\text{i)} \quad \sigma = (1,2)$$
$$\text{ii)} \quad \sigma = (1,2,\ldots,n+3)$$

as these generate S_{n+3}.

For $\sigma = (1,2)$,

$$\begin{aligned} a_1a_2a_3\ldots a_{n+2}a_{n+3} &= a_1(a_2(a_3\ldots a_{n+2})a_{n+3}) \\ &= a_1(a_{n+3}(a_3\ldots a_{n+2})a_2) = a_1((a_{n+3}a_3)\ldots a_{n+2})a_2 \\ &= a_2((a_{n+3}a_3)\ldots a_{n+2})a_1 = a_2(a_{n+3}(a_3\ldots a_{n+2})a_1) \\ &= a_2(a_1(a_3\ldots a_{n+2})a_{n+3}) = a_2a_1a_3\ldots a_{n+2}a_{n+3}. \end{aligned}$$

For $\sigma = (1,2,\ldots,n+3)$,

$$\begin{aligned} a_1a_2a_3\ldots a_{n+2}a_{n+3} &= a_1(a_2(a_3\ldots a_{n+2})a_{n+3}) \\ &= a_1(a_{n+3}(a_2\ldots a_{n+2})a_2) = a_1a_{n+3}(a_3\ldots a_{n+2})a_2 \end{aligned}$$

and by the previous part just proved,

$$a_1a_2a_3\ldots a_{n+2}a_{n+3} = a_{n+3}a_1(a_3\ldots a_{n+2})a_2 = a_2(a_3\ldots a_{n+2})a_{n+3}a_1. \;∎$$

<u>Corollary</u>. *Let* F_2V_n *be the* V_n*-free algebra of rank* 2 *generated by* $\{a,b\}$. *If* $m \geq 3$, *then* $a^m b^n = b^n a^m$. *Moreover, if* $a_1a_2\ldots a_{n+m} \in F_2V_n$, *then* $a_1a_2\ldots a_{n+m} = a^k b^\ell$ *for some natural numbers* k *and* ℓ. ∎

<u>Lemma</u>. $V_n \subset V_{n+1}$ *for all* n.

Proof. The inclusion is obvious. Therefore, we shall only give a model of V_{n+1} which does not belong to V_n. Let A be any set (i.e. "alphabet") with $|A| > 1$, and let P be the set of nonempty words in A, multiplied in the usual way by writing two words together. The resulting algebra $(P,*)$ is the free semigroup generated by A. Define a congruence θ on $(P,*)$ as follows:

$(w_1,w_2) \in \theta$ iff length w_1 = length $w_2 \geq n+3$ and w_2 is a rearrangement of w_1.

Obviously $(P,*)/\theta \in V_{n+1}$, because the law $x(y_1 \dots y_{n+1})z \approx z(y_1 \dots y_{n+1})x$ involves two words of length n+3 or more. However, $a \cdot a^n \cdot b \neq b \cdot a^n \cdot a$ in $(P,*)/\theta$ and hence $(P,*)/\theta \notin V_n$. ∎

Proposition 73.

i) $V_{n+1} < V_n$ *for all* n.

ii) Semc < Semℓ.

Proof. Obvious by applying Proposition 72. ∎

Corollary.

$$W < V_n.$$

Proof. Is is sufficient to show that $W \leq V_{n+1}$. Define $x \circ y = x^{n+1}y^{n+1}$. This is a W-operation, because $x \circ y \approx x^{n+1}y^{n+1} = y^{n+1}x^{n+1} \approx y \circ x$.

$$(x \circ y)\circ(u \circ v) \approx (x^{n+1}y^{n+1})^{n+1}(u^{n+1}v^{n+1})^{n+1} \approx x^{(n+1)^2}u^{(n+1)^2}y^{(n+1)^2}v^{(n+1)^2}$$
$$\approx (x \circ u)\circ(y \circ v)$$

Note the use of the corollary of our previous lemma to obtain both of these equations. ∎

Let C be the variety of constant semigroups which satisfy $xy \approx zt$ and $(xy)z \approx x(yz)$. (Already mentioned in Proposition 72).

Proposition 74. Semc < C

Proof. Since constant semigroups are commutative, C is a subvariety of Semc. Therefore Semc $\leq$ C.

Moreover, no term of F_2Semc defines a C-operation. In fact, only terms $a^n b^n$ can define a commutative operation $xw_n y = x^n y^n$, and this one is not constant. (Example in $(\mathbb{Z},+)$, $1w_n 1 = 2n \neq n = 1w_n 0$).

Therefore Semc < C. ∎

Let Bin1 be the variety of type $\langle 2,0 \rangle$ defined by $1 \cdot x = x = x \cdot 1$ where 1 is its nullary operation. This variety has of course been very important in the development of algebraic topology. Its models in the category of topological spaces and continuous maps (sometimes the category of spaces and homotopy classes of maps) are known as H-spaces. Thus the spaces which can be given the structure of an H-space are precisely the spaces A with $V(A) \geq$ Bin1 in our lattice L. (For V(A) see §1.)

Since there exist spaces which cannot be H-spaces, but of course any space admits a constant operation, the next proposition will be completely evident to those familiar with these topological ideas.

Proposition 75. $C < \text{Bin } 1$

Proof. The forgetful functor $F : \text{Bin } 1 \to C$

$$(M,\cdot,1) \to (M,1)$$

yields immediately $C \leq \text{Bin } 1$.

Now $F_2C = \{1,a,b\}$ has only three terms, none of which gives a Bin1-operation:

$x \circ y = 1$ would lead for algebras with more than one element to $a \circ 1 = 1 \neq a$. Moreover, $x \circ y = x$ leads to $1 \circ a = a \neq a$ and $x \circ y = y$ leads to $a \circ 1 = 1 \neq a$. ∎

In fact, as we shall see in §11, there is a proper class of varieties between C and Bin1.

Proposition 76. Mon < Gru.

Proof. A forgetful functor yields Mon $\leq$ Gru. Let M be the following 5-element monoid:

$\cdot$	1	a	b	c	d
1	1	a	b	c	d
a	a	a	a	a	a
b	b	b	b	b	b
c	c	c	c	c	c
d	d	d	d	d	d

Associativity is checked by:

If $x \neq 1$, then $(xy)z = xz = x(yz)$.

If $x = 1$, then $(xy)z = yz = x(yz)$.

Now $\text{Aut}(M) \cong S_4$ and the only groups G which can be defined on $\{1,a,b,c,d\}$ are isomorphic to $\mathbb{Z}_5$ and have therefore only 4 automorphisms. Therefore, $\text{File}(M) \not\geq \text{File}(G)$. ∎

Proposition 77.

i) Bin 1 $<$ Mon

ii) Mon $<$ Monc

Proof. Obvious by Proposition 72. ∎

The 7-dimensional sphere S^7 can be made an H-space (see the discussion after Proposition 74) by multiplying Cayley numbers of unit modulus, and thus $V(S^7) \geq$ Bin1. But Adams' celebrated result that S^7 cannot be an *associative* H-space tells us that $V(S^7) \not\geq$ Mon. Thus we have a second (albeit unnecessarily sophisticated) proof that Bin1 $<$ Mon. ∎

Proposition 78. Monc $<$ Ab.

Proof. As Monc $\leq$ Ab is obvious, we consider the following algebra B,

	1	a	b	c	d
1	1	a	b	c	d
a	a	1	b	c	d
b	b	b	b	c	d
c	c	c	c	c	d
d	d	d	d	d	d

and its subalgebra A,

	1	a
1	1	a
a	a	1

Note that B can be obtained from A by successively adding three zeros. Therefore $A \to B$ is a Monc homomorphism. Since no group G with five elements has a subgroup with two, File(B) $\not\geq$ File(G). ∎

(Thus, in fact, Monc and Ab are separated by the "Lagrangian" filter of Taylor [72], p. 380.)

Proposition 79. *Let* $\mathcal{U}_n$ *be the variety with only one binary operation, satisfying* $x^n \cdot y^n = y^n \cdot x^n$. *Then* $\mathcal{U} =_L \mathcal{U}_n$ *for all natural* n.

Proof. Clearly $\mathcal{U} = \mathcal{U}_1$, and $\mathcal{U}_1$ is a subvariety of $\mathcal{U}_n$. Therefore $\mathcal{U}_n \leq \mathcal{U}$. Now define the following derived operation of $\mathcal{U}_n$: $x \circ y := x^n \cdot y^n$. Since $x \circ y = y \circ x$, $\circ$ is a $\mathcal{U}$-operation and $\mathcal{U} \leq \mathcal{U}_n$ is shown. ∎

Let $\mathcal{U}_n^{as}$ be the subvariety of $\mathcal{U}_n$ for which the operation is associative. It is obvious that $\mathcal{U}_n^{as} \subset W$; e.g. $(\mathbb{Z}, 2x+2y) \in W - \mathcal{U}_n^{as}$.

Proposition 80.

$$W < \mathcal{U}_n^{as}$$

Proof. Since W is the class of models of $xy = yx$ and $(xy)(uv) = (xu)(yv)$ one easily sees, as in the Corollary to Proposition 73, that the derived operation $x \circ y = x^n y^n$ is a W-operation, and $W \leq \mathcal{U}_n^{as}$ follows.

A slightly sharpened version of Proposition 72, which includes the case $A = W$, and has a very similar proof, completes our present argument.∎

Corollary. $W < \text{Semc}$.

Proof. $\mathcal{U}_n^{as} \leq \text{Semc}$ as Semc is a subvariety.∎

Lemma.

i) $V_{2n-3} \subset \mathcal{U}_n^{as}$ for $n > 1$.

ii) *If* $m \neq n$ *and* $m|n$, *then* $\mathcal{U}_m^{as} \subset \mathcal{U}_n^{as}$.

iii) *If* $m \nmid n$ *and* $n \nmid m$, *then* $\mathcal{U}_n^{as} \not\subseteq \mathcal{U}_m^{as}$ *and* $\mathcal{U}_m^{as} \not\subseteq \mathcal{U}_n^{as}$.

iv) *If* $m \geq 2n-2$, *and* $n \geq 2$, *then* $V_m \not\subseteq \mathcal{U}_n^{as}$ *and* $\mathcal{U}_n^{as} \not\subseteq V_m$.

Proof. For every n, define on $F_2\text{Sem}$ a congruence θ_n by identifying two words ω_1 and ω_2 if and only if one of the following four cases happens:

(1) $\omega_1 = p^n q^n$ and $\omega_2 = q^n p^n$ for some $p,q \in F_2\text{Sem}$

(2) $\omega_1 = up^n q^n$ and $\omega_2 = uq^n p^n$ for some $p,q,u \in F_2\text{Sem}$

(3) $\omega_1 = p^n q^n v$ and $\omega_2 = q^n p^n v$ for some $p,q,v \in F_2\text{Sem}$

(4) $\omega_1 = up^n q^n v$ and $\omega_2 = uq^n p^n v$ for some $u,p,q,v \in F_2\text{Sem}$.

Then $F_2\text{Sem}/\theta_n \in \mathcal{U}_n^{as}$.

To prove i), $V_{2n-3} \subset \mathcal{U}_n^{as}$ follows by the Corollary preceding Proposition 73. Moreover

$$a\cdot b^{2n-3}\cdot b = a\cdot b^{2n-2} \neq b^{2n-2}\cdot a = b\cdot b^{2n-3}\cdot a$$

in $F_2\text{Sem}/\theta_n$ and therefore $F_2\text{Sem}/\theta_n \notin V_{2n-3}$.

Now let $n = md$ with $d > 1$. Then

$$a^n b^n = (a^d)^m\cdot(b^d)^m = (b^d)^m(a^d)^m = b^n a^n$$

and $\mathcal{U}_m^{as} \subset \mathcal{U}_n^{as}$ follows.

But since $m < n$, $a^m\cdot b^m \neq b^m\cdot a^m$ in $F_2\text{Sem}/\theta_n$ and one obtains $F_2\text{Sem}/\theta_n \in \mathcal{U}_n^{as}$ and $F_2\text{Sem}/\theta_n \notin \mathcal{U}_m^{as}$, which yields ii).

Assume that $m \nmid n$ and $n \nmid m$. Without loss of generality, let $m > n$. From our preceding argument, we already have $\mathcal{U}_m^{as} \not\subseteq \mathcal{U}_n^{as}$. Since $m > n$ and $n \nmid m$, there are two integers d and q, $0 < q < n$ such that $m = nd+q$. Then

$$a^m b^m = a^q (a^d)^n (b^d)^n b^q = a^q b^{dn} a^{dn} b^q$$

$$\neq b^q b^{dn} a^{dn} a^q = b^m a^m \quad \text{in } F_2\text{Sem}/\theta_n.$$

Therefore $F_2\text{Sem}/\theta_n \notin U_m^{as}$ and $U_n^{as} \not\subseteq U_m^{as}$ is obtained.

To show iv), Let $m > 2n-2$. Call $\theta_{(m)}$ the congruence defined on $F_2\text{Sem}$ by

$(w_1,w_2) \in \theta_{(m)}$ iff length w_1 = length $w_2 \geq m+2$ and w_2 is a rearrangement of w_1.

Then $F_2\text{Sem}/\theta_{(m)} \in V_m$ because the law

$$xy_1\ldots y_m z \approx zy_1\ldots y_m x$$

involves two words of length $m+2$ or longer.

Since $a^n \cdot b^n \neq b^n a^n$ in $F_2\text{Sem}/\theta_{(m)}$ this semigroup does not belong to U_n^{as} and $V_m \not\subseteq U_n^{as}$ follows. But also, since $a \cdot b^m \cdot b \neq b \cdot b^m \cdot a$ in $F_2\text{Sem}/\theta_n$, $U_n^{as} \not\subseteq V_m$ also follows. ∎

<u>Proposition 81</u>.

i) *If* $m \neq n$ *and* $m|n$, *then* $U_n^{as} < U_m^{as}$.

ii) *If* $m \nmid n$ *and* $n \nmid m$, *then* U_n^{as} *and* U_m^{as} *are incomparable in* L.

iii) $U_n^{as} < V_{2n-3}$ *for* $n > 1$.

iv) *If* $m > 2n-2$ *and* $n \geq 2$, *then* U_n^{as} *and* V_m *are incomparable in* L.

<u>Proof</u>. Follows from the lemma just proved and Proposition 72. ∎

<u>Corollary</u>. *The breadth of* [W,Ab<u>Sem</u>] *in* L *is infinite.* ∎

We call U_n^{as} the variety of type $\langle 2 \rangle$ which consists of the models of the following equations:

$$x^n y^n \approx y^n x^n$$

$$(xy)(uv) \approx (xu)(yv)$$

<u>Proposition 82</u>. *For all natural* n,

$$W =_L U_n^{was}$$

Proof. Clearly $\mathcal{U}_n^{was} \leq W$ as W is a subvariety. Since

$$(x^n y^n)^{m+1} \approx (x^{nm} y^{nm})(x^n y^n) \approx (x^{nm} x^n)(y^{nm} y^n) \approx x^{n(m+1)} y^{n(m+1)}$$

with the parentheses in this non-associative variety properly understood, (i.e., $x^4 = ((xx)x)x$ and $x^{n\cdot 3} = (x^n x^n)x^n$). By induction, $(x^n y^n)^m = x^{nm} y^{nm}$.

Define $x \circ y = x^n y^n$ a derived operation in $\mathcal{U}_n^{was}$. Now $x \circ y$ is clearly commutative, and besides:

$$(x \circ y) \circ (u \circ v) = (x^n y^n)^n (u^n v^n)^n = (x^{nn} y^{nn})(u^{nn} v^{nn})$$

$$= (x^{nn} u^{nn})(y^{nn} v^{nn}) = (x \circ u) \circ (y \circ v).$$

This shows that $\circ$ is a W-operation, and $W \leq \mathcal{U}_n^{was}$ follows. ∎

Remark. By the Corollary to Proposition 73, we know that $W < \mathcal{V}_n$ and therefore $W \leq \bigwedge_{\mathbb{N}} \mathcal{V}_n$. However, all $\mathcal{V}_n$ are subvarieties of Sem, and $HSP(\bigcup_{\mathbb{N}} \mathcal{V}_n) =$ Sem as can be shown since $\Pi_{\mathbb{N}} F_2 \mathcal{V}_n$ has a subalgebra isomorphic to F_2Sem (generated by the constants $\bar{a}$ and $\bar{b}$).

Before we concentrate on the interval [Sem,W] of L, we shall show that [Sem,C] is a proper class.

Proposition 83. *There is a proper class below* C.

Proof. For every cardinal κ define the variety Q_κ by the following identities:

$$f_i(x,x) = x \qquad \text{for all } i < \kappa$$

$$\lambda_{ij}(f_i(f_j(x,y),f_j(y,x)) = \lambda_{ij}(y) \qquad \text{for all } i < j < \kappa.$$

Obviously $Q_\kappa \leq Q_\lambda \leq C$ for $\kappa \leq \lambda$. We will now prove that if $2^\kappa < \lambda$, then $Q_{2^\kappa} < Q_\lambda$. To begin, we define a model $\mathfrak{A}$ of Q_{2^κ}, whose underlying set is ${}^\kappa 2$, the set of all functions $\sigma : \kappa \to 2$. We let $\mathcal{U}_i$ $(i < 2^\kappa)$ be an enumeration of all subsets of κ. For $\alpha < \kappa$ define

$$f_i(x,y)\ (\alpha) = \begin{cases} x(\alpha) & \text{if } \alpha \in \mathcal{U}_i \\ y(\alpha) & \text{if } \alpha \notin \mathcal{U}_i. \end{cases}$$

For each $i < j$, choose $\tau_{ij} \in \mathcal{U}_i + \mathcal{U}_j$ (Boolean sum) and then define

$$\lambda_{ij}(x)\ (\alpha) = x(\tau_{ij}).$$

We now check the identities. There are two cases: $\tau_{ij} \in \mathcal{U}_i - \mathcal{U}_j$ and $\tau_{ij} \in \mathcal{U}_j - \mathcal{U}_i$; we do only the first one. Obviously $f_i(x,x) = x$. Next

$$\begin{aligned}\lambda_{ij}(f_i(f_j(x,y),f_j(y,x))\ (\alpha) \\ = f_i(f_j(x,y),f_j(y,x))\ (\tau_{ij}) \\ = f_j(x,y)\ (\tau_{ij}) \\ = y(\tau_{ij}) = \lambda_{ij}(y)\ (\alpha)\end{aligned}$$

and hence $\lambda_{ij}(f_i(f_j(x,y),f_j(y,x)) = \lambda_{ij}(y)$ in $\mathfrak{A}$.

Suppose there is a concrete functor $F : Q_{2^\kappa} \to Q_\lambda$. Since $\lambda > 2^\kappa$, two operations f_i, f_j of Q_λ would have to be interpreted by the *same* derived operation f of $\mathfrak{A}$. Thus we would have f binary and λ unary derived operations of $\mathfrak{A}$ satisfying

$$\text{(a)}\quad f(x,x) \approx x$$

$$\text{(b)}\quad \lambda f(f(x,y),f(y,x)) \approx \lambda y$$

Now f must have the form

$$f(x_1,x_2)\ (\alpha) = x_m(\tau) \quad \text{for } m = 1 \text{ or } m = 2.$$

where τ and m both depend on α but not on x_1,x_2. (Clone($\mathfrak{A}$) consists entirely of such operations). Idempotence (a) immediately assures us that $\tau = \alpha$, and from this we deduce that this operation satisfies

$$f(f(x,y),f(y,x)) \approx f(x,x).$$

From (b) we now deduce the following identity for $\lambda : \lambda(y) = \lambda(x)$.

Hence, λ is a constant function, but this is impossible since $\mathfrak{A}$ has the disjoint subalgebras $\{(0,0,\dots)\}$ and $\{(1,1,\dots)\}$. ∎

Remark. Recalling the varieties W_κ defined after Proposition 41, since $W_{2^\kappa} = HSP\{\mathfrak{A}\}$ we have

$$Q_{2^\kappa} \leq W_{2^\kappa} \quad \text{but}$$

$$Q_\lambda \nleq W_{2^\kappa} \quad \text{for } \lambda > 2^\kappa.$$

Since of course also $Q_\lambda < W_\lambda$ we have an alternate proof that $W_\lambda \nleq W_{2^\kappa}$ for $\lambda > 2^\kappa$.

From now on we shall consider the interval [Sem,W] of L. Recall from our notation on varieties that X_n means the models of

$$x_1 \dots x_{n-1} x_n \approx x_n(x_1 \dots x_{n-1})$$

in type $\langle 2 \rangle$, where we write for simplicity $x_1x_2x_3\dots x_n$ instead of $(\dots(((x_1 \cdot x_2)x_3)\dots)x_n$.

The variety Y of the same type is defined by

$$x_1x_2x_3 \approx x_3(x_1x_2) \quad \text{and} \quad x_1x_2x_3 \approx x_2x_1x_3$$

and is a cover of $\mathcal{U}$ in the lattice of varieties of type $\langle 2 \rangle$. (See [30].) Unfortunately it is *not* a cover in L as we see next.

Proposition 84. $\mathcal{U} =_L X_n = Y$ *for all* $n > 1$.

Proof. We first notice that if $n < m$, X_n is a subvariety of X_m and $X_2 = \mathcal{U}$. Let $n < m$ and $\tau = m-n$. Let $y = x_1 \ldots x_m$ and $z = x_1 \ldots x_{r+1}$. Then $y = zx_{r+2} \ldots x_m = x_m(zx_{r+2} \ldots x_{m-1}) = x_m(x_1 \ldots x_{m-1})$. Therefore $X_n \quad X_m$. Moreover, we also claim that X_{n+1} is a subvariety of $\mathcal{U}_n$. Indeed, if $\mathfrak{A} \varepsilon X_{n+1}$, it satisfies $a^n b^n = (a \ldots a)b^n = b^n(a \ldots a) = b^n a^n$, and $\mathfrak{A} \varepsilon \mathcal{U}_n$.

We have obtained the following chain of subvarieties:

$$\mathcal{U} = X_2 \subset X_{n+1} \subset \mathcal{U}_n,$$

and translating this into L, and applying Proposition 79,

$$\mathcal{U}_n \leq X_{n+1} \leq X_2 = \mathcal{U} =_L \mathcal{U}_n .$$

On the other hand, since $X_3 \leq Y \leq \mathcal{U}$ and $\mathcal{U} =_L X_3$, $\mathcal{U} =_L Y$ follows. ∎

Proposition 85. *Let* R_n *and* R *be varieties of type* $\langle 2 \rangle$, R_n *the models of*

$$(x_1 \ldots x_n)(x_n \ldots x_1 y) \approx (x_n \ldots x_1 y)(x_1 \ldots x_n)$$

and R *the models of the infinite set of these equations, one for each natural number* n. *Then*

$$\text{Sem} =_L R_1 < R_n \leq R < \mathcal{U} \text{ for } n > 1.$$

Proof. Since S defined by

	a	b
a	a	a
b	b	b

satisfy $x(xy) = (xy)x$, $S \varepsilon R_1$.

Therefore $\text{Sem} =_L R_1$.

Since neither S nor $\overline{S}$ defined by

	a	b
a	a	a
b	a	b

are in any R_n with $n \geq 2$, $\text{Sem} < R_n$.

We show next that $R_n < \mathcal{U}$, since essentially the same proof yields $R < \mathcal{U}$.

To do so, we first remark that in F_2R_n length is well defined, because we only have commutative laws. Moreoever, if $\omega_1\omega_2 = \omega_3\omega_4$, then either $\omega_1 = \omega_3$ and $\omega_2 = \omega_4$ or $\omega_1 = \omega_4$ and $\omega_2 = \omega_3$, and whenever $\omega_1 \neq \omega_2$ and $\omega_1\omega_2 = \omega_2\omega_1$, then length $\omega_1 \neq$ length ω_2. These remarks are three very easy statements which can be proved by induction.

To convince ourselves that $R_n < \mathcal{U}$, we shall consider the terms of F_2R_n and show that none of them defines a commutative derived operation. This statement is proved by induction on the length of the terms.

Let τ be the automorphism of F_2R_n defined by the permutation (a,b). A necessary and sufficient condition for a word to define a commutative operation is $\tau(\omega) = \omega$. If length $\omega = 1$, $\omega \in \{a,b\}$, and so ω obviously does not define a commutative operation.

Since the type is $\langle 2 \rangle$, i.e. we have only one operation which is binary, if length $\omega > 1$ there are two words ω_0, ω_1 of shorter length such that $\omega = \omega_0\omega_1$. By our induction hypothesis, no word of length less than k defines a commutative derived operation. Let ω be a word of length k, $k > 1$. Then there exist two words ω_0, ω_1 of length less than k such that $\omega = \omega_0\omega_1$.

Suppose ω defines a commutative derived operation. Then $\omega_0\omega_1 = \omega = \tau(\omega) = \tau(\omega_0)\tau(\omega_1)$, and since $\omega_0 \neq \tau(\omega_0)$ and $\omega_1 \neq \tau(\omega_1)$, by our previous remarks, $\omega_0 = \tau(\omega_1)$ and $\omega_1 = \tau(\omega_0)$. But then $\omega_0\omega_1 = \omega_1\omega_0$ and by the same set of remarks, length $\omega_0 \neq$ length ω_1, a fact which contradicts $\omega_0 = \tau(\omega_1)$. ▮

Clearly the whole argument applies also for F_2R.

Corollary. *Let* T *be a non-empty finite set of natural numbers. Then* $0 < \bigwedge_{t \in T} R_n$.

Proof. Follows from last Proposition and the fact that 0 is a $\bigwedge$-prime element of L, as shown in the Corollary to Proposition 18. ▮

Problem. $0 = \bigwedge_{\mathbb{N}} R_n$?

Proposition 86. *Let* m,n *be distinct natural numbers larger than* 2. *Then*

i) R_m *and* R_n *are incomparable*

ii) $R_m < R$

iii) $R_n \nleq R_2$

iv) $R_2 < R$.

Proof. For "i)" by symmetry, it will suffice to show that $R_n \nleq R_m$. To see that no term of F_2R_m can define a R_n-operation, suppose $\omega(a,b)$ defines a R_n-operation $x \circ y = \omega(x,y)$. Then

$$t_1 = (x \circ y \circ x_1 \circ \dots \circ x_{n-2}) \circ (x_{n-2} \circ \dots \circ x_1 \circ y \circ x \circ u) =$$

$$(x_{n-2} \circ \dots \circ x_1 \circ y \circ x \circ u) \circ (x \circ y \circ x_1 \circ \dots \circ x_{n-2}) = t_2$$

Since length is well-defined in F_2R_m even on each variable, if $\text{length}_x \omega(x,y) = \ell$ and $\text{length}_y \omega(x,y) = k$, $\text{length}_x t_1 = \ell^n + k^2\ell$ and $\text{length}_x t_2 = k\ell^2 + \ell^{n-1}$. Moreover, $\text{length}_u t_1 = k^2$ and $\text{length}_u t_2 = k\ell$.

Therefore $k^2 = k\ell$, and for $k \neq 0$, $k = \ell$, and also $\ell^n + k^2\ell = k\ell^2 + \ell^{n-1}$, which combined with the above result, gives $\ell^n = \ell^{n-1}$, and $\ell = 1 = k$.

We are left with the terms ab and ba and none of them satisfy $t_1 = t_2$.

For "ii)" we know from Proposition 85 that $R_m \leq R$, and $R \nleq R_m$ follows from the same discussion as in "i)" above, since an R-operation still should satisfy $t_1 = t_2$ of i).

Still the same argument proves iii) and iv). ∎

Corollary. *If* m,n *are distinct integers larger than two, then* $R_m \wedge R_n < R_n$ *and* $R_n \wedge R_2 < R_n$. ∎

Corollary. *The breadth of the interval* [Sem,$\mathcal{U}$] *in* L *is infinite.* ∎

The variety V_4 does not cover V_2, because $2 \in \operatorname{Spec} V_2 - \operatorname{Spec}(V_4 \wedge V_8)$ and $8 \in \operatorname{Spec}(V_4 \wedge V_8) - \operatorname{Spec} V_4$. Therefore $V_2 < V_4 \wedge V_8 < V_4$.

Recall the variety V(A) defined in §2 for any object in a category $\mathcal{C}$. We direct our attention specially to $\mathcal{C}$ being topological spaces. According with Proposition 11, a variety V has a model based on A if and only if $V \leq V(A)$. As examples of this, if we denote by •—• the subspace $[0,1] = \{x \in \mathbb{R} \mid 0 \leq x \leq 1\}$ of $\mathbb{R}$ with its natural topology one easily sees that •—• has also a natural order which makes it into a chain and therefore a topological distributive lattice.

Proposition 87. $\quad \mathrm{Ld} < V(\,\bullet\!\!-\!\!\bullet\,)$

Proof. The above remark gives $\leq$. To see the strict inequality, consider the following polynomial which is a continuous function $\mathbb{R}^3 \to \mathbb{R}$:

$$p(x,y,z) = [(x-y+z) \wedge 1] \vee 0$$

This function is also well-defined in $[0,1]^3$ and its image is also in $[0,1]$.

Since $p(x,x,y) = [(x-x+y) \wedge 1] \vee 0 = (y \wedge 1) \vee 0 = y$ and $p(y,x,x) = [(y-x+x) \wedge 1] \vee 0 = y$, one obtains

$$Mt_1 \leq V(\,\bullet\!\!-\!\!\bullet\,), \qquad \text{while } Mt_1 \nleq \mathrm{Ld}. \ ∎$$

Corollary.

$$Mt_1 < V(\,\bullet\!\!-\!\!\bullet\,)$$

Proof. Suppose $Mt_1 = V(\,\bullet\!\!-\!\!\bullet\,)$. We obtain $Mt_0 < \mathrm{Ld} < V(\,\bullet\!\!-\!\!\bullet\,) = Mt_1$ which contradicts Proposition 70. ∎

Similarly, if we consider the group $(\mathbb{R},+)/\mathbb{Z}$ it can be modeled on an interval $[0,1]$ with 0 and 1 identified. This gives an Abelian group structure to a circle. We denote such circle by 0 and have

Corollary.

$$Ab \leq V(0).\blacksquare$$

We already mentioned after Proposition 77 that also $Bin\ 1 \leq V(S^7)$.

Concerning the kth-power varieties defined after Proposition 5, we show next that $Set^{[2]}$ is not only strictly higher than 0 but also outside the ideal $\mathcal{U}$ of essentially unary varieties. See page 44.

Proposition 88. *The variety* $Set^{[2]}$ *is strictly higher than* 0 *and outside the ideal of essentially unary varieties.*

Proof. Since $\text{Spec } Set^{[2]} = \{n^2 | n < \omega\}$, $0 < Set^{[2]}$ follows immediately. From its definition, the identities of $Set^{[2]}$ are:

1) $f(f(x)) \approx x$

2) $x^2 \approx x$

3) $(xy)(uv) \approx xv$

4) $f(x \cdot y) \approx f(y) \cdot f(x)$

Suppose $Set^{[2]}$ is essentially unary. Let V be a unary variety such that $Set^{[2]} \leq V \leq 1$. There is a concrete functor $F : V \to Set^{[2]}$. The only terms in $F_2 V$ are $g(x)$ and $g(y)$. Without loss of generality, $x \cdot y = g(x)$. From 2), we deduce $g(x) \approx x \cdot x \approx x$ and therefore $x \cdot y \approx x$. From 4), $f(x) \approx f(xy) \approx f(y) \cdot f(x) \approx f(y)$, and from 1), $x \approx ffx \approx ffy \approx y$. Since only algebras with one element are interpreted by F, we conclude $V = 1$. A contradiction.$\blacksquare$

11. A PROPER CLASS BETWEEN C AND BIN 1

The authors are indebted to J. Mycielski for having suggested the study of the varieties V_{a_α} which we introduce as follows. If α is a cardinal number, we consider the type with one 4-ary operation f and one 0-ary operation c_ξ for each $\xi < \alpha$. We call V_{a_α} the variety of models of the equations

$$f(x,y,c_\xi,c_\eta) = x \quad \text{if } \xi = \eta$$

$$f(x,y,c_\xi,c_\eta) = y \quad \text{if } \xi \neq \eta.$$

Lemma. *If* $1 < \beta < \alpha$, *then* $V_{a_\alpha} \nleq V_\beta$.

Proof. Let $\beta < \alpha$ and suppose that $V_{a_\alpha} \leq V_\beta$. There is a concrete functor $F : V_\beta \to V_{a_\alpha}$.

Since every constant of V_{a_α} will be interpreted by a term in F_0V_β, we have at most $\beta = |F_0V_\beta|$ different constants in F_0V_β.

Therefore, there is $\lambda < \mu < \alpha$ such that $c_\lambda = c_\mu$. Then $y = f(x,y,c_\lambda,c_\mu) = f(x,y,c_\lambda,c_\lambda) = x$ in $F_2V_{a_\alpha}$. Since $1 < \alpha = |F_2V_{a_\alpha}|$, this is a contradiction. ∎

Proposition 89. *The supremum of the class* $(V_{a_\alpha})_{\alpha \in \mathrm{Card}}$ *is* 1.

Proof. Let $V \neq 1$ and $M \in V$ such that $1 = \lambda = |M|$. Without loss of generality λ is infinite. Then $V \leq V_\lambda$. Suppose $V_{a_{2^\lambda}} \leq V$. Then $V_{a_{2^\lambda}} \leq V_\lambda$ which contradicts the lemma. ∎

Proposition 90. *If* $1 < \beta < \alpha$, *then* $V_{a_\beta} < V_{a_\alpha}$.

Proof. Since V_{a_α} satisfies the equations of V_{a_β}, plus additional ones, a forgetful functor deleting c_ξ for $\beta \leq \xi < \alpha$ yields $V_{a_\beta} \leq V_{a_\alpha}$.

If we assume $V_{a_\alpha} \leq V_{a_\beta}$, the same argument as in the lemma before Proposition 89 involving the cardinality of $F_0V_{a_\beta}$ will yield the contradiction. ∎

Proposition 91. $\mathrm{Bin}\ 1 \nleq V_{a_\alpha}$

Proof. For each $\eta < \alpha$, let $(A_\eta,+)$ be a *non-commutative* group. Assume that if $y \neq y'$, then $A_y \cap A_{y'} = 0$.

Define $A = \bigcup_{\gamma<\alpha} A_\gamma$.

Now give A the structure of a V_{a_α}-algebra as follows:

c_η the identity of the group $(A_\eta,+)$

$$f(x,y,z,w) = \begin{cases} x & \text{if } z,w \in A_\xi \text{ for a } \xi < \alpha \\ y & \text{otherwise} \end{cases}$$

It is clear that $(A,c_\eta,f) \in V_{a_\alpha}$.

Define $G \subseteq A^3$ as follows:

$$G = \bigcup_{\eta<\alpha} \{(x,y,z) \in A_\eta^3 \mid x + y = z\}.$$

G is clearly a subalgebra of A^3.

Now suppose there were a concrete functor $V_{a_\alpha} \to \text{Bin } 1$. This means that A has a binary derived operation $*$ and a constant e such that

1) $(A,*,e) \in$ Bin 1 and

2) G is a subalgebra of $(A,*,e)^3$.

From 1) and 2) we will derive a contradiction. Suppose $e \in A_\eta$. Since G is a subalgebra, we have $(e,e,e) \in G$ which yields $e + e = e$ in $(A_\eta,+)$ and means that e is the identity of $(A_\eta,+)$. Now select any $x,y \in A_\eta$. Since $(A_\eta,+,e)$ is a group, we have

$$\left.\begin{matrix}(x,e,x)\\(e,x,x)\\(y,e,y)\\(e,y,y)\end{matrix}\right\} \in G.$$

Since G is a $*$-subalgebra, it also contains

$$(x,e,x)*(e,y,y) = (x*e,e*y,x*y) = (x,y,x*y)$$

and

$$(e,x,x)*(y,e,y) = (e*y,x*e,x*y) = (y,x,x*y).$$

By definition of G, we have

$$x + y = x*y = y + x,$$

in contradiction to our assumption that $(A_\eta,+)$ is non-commutative. ∎

<u>Proposition 92</u>. *If* $\alpha < 2$, *then* $V_{a_\alpha} \not\leq$ Bin 1.

<u>Proof</u>. Bin 1 has a model with 2 elements while every non-trivial model of V_{a_α} has at least α elements. ∎

<u>Corollary</u>. $V_{a_\alpha} \| \text{Bin } 1$ *for all* $\alpha > 2$. ∎

Proposition 93. *Let* B,V,W *be varieties such that for every finite* n, *if* $\mathfrak{A} \varepsilon$ V *and* $|\mathfrak{A}| \,\big|\, |F_n W|$, *then* $|\mathfrak{A}| = 1$. *If* B $\not\leq$ W, *then* B $\wedge$ V $\not\leq$ W.

Proof. Suppose there were a concrete functor $\phi : W \to B \times V$. We shall show that $\phi(W) \subseteq B$ leading to another concrete functor also denoted by $\phi : W \to B$ which is a contradiction.

Since W is generated by its free algebras of finite rank $F_n W$, we consider $\phi(F_n W) = \mathfrak{B}_n \times \mathfrak{A}_n$ with $\mathfrak{B}_n \varepsilon$ B and $\mathfrak{A}_n \varepsilon$ V. Now $|\mathfrak{A}_n| \,\big|\, |\mathfrak{B}_n \times \mathfrak{A}_n| = |F_n W|$ and by our hypothesis $|\mathfrak{A}_n| = 1$. Thus $\phi(F_n W) \cong \mathfrak{B}_n \varepsilon$ B. ∎

In order to obtain a proper class in the promised location, we now simply define $V_{ab_\alpha} = V_{a_\alpha} \wedge$ Bin 1 for each cardinal α.

Corollary. *If* $\aleph_0 \leq \beta < \alpha$, *then* $V_{ab_\beta} < V_{ab_\alpha}$.

Proof. Since $V_{a_\beta} < V_{a_\alpha}$, $V_{ab_\beta} \leq V_{ab_\alpha}$ follows. Since Bin 1 $\not\leq V_{a_\beta}$ and $\aleph_0 \leq \beta < \alpha$, $|F_n V_{a_\beta}| = \beta$ for all finite n. By Proposition 91, V_{ab_α} = Bin 1 $\wedge V_{a_\alpha} \not\leq V_{a_\beta}$. Therefore $V_{ab_\alpha} \not\leq V_{ab_\beta} \leq V_{a_\beta}$. ∎

This corollary shows that there is a proper class of varieties, even a chain, between C and Bin 1.

A further application of Proposition 93 will also show that the varieties {Lat $\wedge$ Ab_p | p prime} and even {$Mt_0 \wedge Ab_p$ | p prime} are all distinct in L.

One also obtains $W_k \wedge C < W_\lambda \wedge C$ for $k < \lambda$ and k a power of 2. (This is essentially the same as the proper class Q_k from Proposition 83.)

OPEN PROBLEMS

In addition to the following problems, which relate to all phases of this research, we also refer the reader to some problems which have been more natural to state at the appropriate places in the main body of the Memoir. See pages 31, 35, 37, 49, 58, 66, 91, and especially the long sequence of problems on the special group-theoretic aspects of this subject at the end of §9 (pages 94-96).

Problem 1. Find a cover of 0 or an infinite family of varieties $V_n > 0$ such that $\bigwedge V_n = 0$.

Problem 2. Find non-trivial covers.

Problem 3. Find a proper class of varieties $\mathcal{V}_\lambda$, where $\lambda < \mu$ implies $\mathcal{V}_\mu < \mathcal{V}_\lambda$ for cardinals λ, μ.

Problem 4. Are any of the subintervals denoted by lines in the pictures of finite (or bounded) length? Which ones contain a proper class?

Problem 5. Is the breadth of L uncountable? Is there an antichain which is a proper class?

Problem 6. Is there a variety $\mathcal{V}$ such that $L = [0,\mathcal{V}] \cup [\mathcal{V},1]$? Is there a minimal antichain if the answer of the above is negative?

Problem 7. Does every subclass of L have an inf and a sup?

Problem 8. Is $MQ1 < Qgr$?

Problem 9. Define $MQ2$ and $SMQ2$ and decide whether $0 < MQ2 < SMQ2$. Is $MQ2 < Qgr$?

Problem 10. Compare all the varieties in Problems 8 and 9 with Mt_1.

Problem 11. Do a study similar to the one made in §9 for varieties of groups, this time for varieties of semigroups, of monoids, of rings, of modules over special rings.

Problem 12. Subdivide intervals in the pictures by the following procedure. If $A < B$ and $A < C_i$ where $(C_i)_{i \in I}$ is a family of distinct varieties such that $B \| C_i$, obtain $(B \wedge C_i)_{i \in I}$ and locate this family, comparing $B \wedge C_i$ with $B \wedge C_j$.

Problem 13. Can one develop a concept of product of varieties with "amalgamation," which would be relevant in L?

Problem 14. Can one define a "sinality degree" of the algebras in such a way that the concept becomes useful? Say: An algebra A is "2-Sin in V if there are at most two algebras $\mathfrak{A}_1$, $\mathfrak{A}_2$ in V such that if $A \in B$ and $F : B \to V$ is a concrete functor, then $F(A) \cong \mathfrak{A}_1$ of $F(A) \cong \mathfrak{A}_2$.

Could one classify the algebras in any variety by their "n-sinality"? Does n-sinality reflect the degree or complexity of an algebra or does it not?

Problem 15. Are there other "universally SIN-algebras" like $(\mathbb{Z},+)$ in Proposition 72? Can $(\mathbb{Z},+)$ also be useful for non-selfdual varieties? Is a new concept hiding here?

Problem 16. Are there different degrees or qualities of interpretations, depending on categorical properties of the concrete functor involved? Such as: We may have $V_0 < W_0$ and $V_1 < W_1$ but the concrete functor $F_0 : W_0 \to V_0$ is an epimorphism such that $\mathsf{HSP}\, F_0 W_0 = V_0$ while $\mathsf{HSP}\, F_1 W_1 \subset V_1$ for every possible concrete functor $F_1 : W_1 \to V_1$.

Problem 17. Is it important for an interpretation to be "injective"? (For two non-isomorphic algebras $\mathfrak{A}$, $\mathfrak{B}$, $F\mathfrak{A} \ncong F\mathfrak{B}$).

Problem 18. Can one define faithful interpretations? (Something slightly weaker than equivalence of varieties.)

Problem 19. Can one characterize those varieties which are identified in L?

Problem 20. Study the difference between the following three possibilities for $V =_L W$. All these involve concrete functors $F : V \to W$ and $G : W \to V$.

1) Equivalence of varieties: Categorical isomorphism involving concrete functors. ($F \circ G = 1_W$ and $G \circ F = 1_V$.)
2) There exist natural transformations
$$F \circ G \to 1_W \text{ and (or) } G \circ F \to 1_V .$$
3) No clear connection between F and G can be found.

Problem 21. Are some interpretations "better" than others? For example: an abelian group can be interpreted as a ring by defining $x \cdot y = 0$. The corresponding concrete functor $F : Ab \to An$ is very far from being an epimorphism. $HSP\,F\,Ab$ is only a small (perhaps uninteresting) subvariety of An. Could one rank interpretations according to this idea?

Problem 22. (Jan Mycielski) Can one define an adequate concept of interpretation, leading first to a smaller lattice (with many more varieties identified) and later to a graduation of concepts leading to finer study of the varieties?

Problem 23. (Jan Mycielski) Study and locate "complete" and "precomplete" varieties.

Problem 24. Is every free lattice a sublattice of L?

Problem 25. Find Whitman failures in L. (See [29].)

Problem 26. Analyze the identification $Ab_n =_L B_n N_3 M$ for $2 \nmid n$ and $3 \nmid n$ in Figure 2. Suppose we define $V \leq W$ if there is an epimorphic concrete functor $F : W \to V$. Are still some of these varieties equal in the new lattice L'?

Problem 27. (Jan Mycielski) Study $\vee$-irreducible and $\vee$-prime elements in the interval $[c,1]$. This problem is motivated by the observation that the join-reduction of $\mathcal{G}$ = groups, etc., seen at the end of §6, proceeds by forming the joinand $\mathcal{G}_0$, which differs very little from $\mathcal{G}$, except in having no constants. A proof that, say, groups form a $\vee$-irreducible variety in $[c,1]$, would support our natural intuition that the variety of groups cannot be built from simpler components.

REFERENCES

1. Bachmuth, S., Mochizuki, H.Y., Walkup, D., A nonsolvable group of exponent 5, Bull. AMS 76 (1970), 638-640.

2. Baker, K.A., Equational classes of modular lattices, Pacific J. Math. 28 (1969), 9-15.

3. Banaschewski, B., On the category invariance of cardinalities in a prevariety, Algebra Universalis 14 (1982), 267-268.

4. Berman, J., Free spectra of 3-element algebras, Proceedings of the Fourth International Conference on Universal Algebra and Lattice Theory, Puebla, 1982, *Universal Algebra and Lattice Theory*, L.N.M. 1004 (1983), 10-53.

5. Bing, R.H., Conditions that a 3-manifold be S^3, Annals of Math. 68 (1978), 17-37.

6. ———— A set is a 3-cell if its Cartesian product with an arc is a 4-cell, AMS Proceedings 12 (1961), 13-19.

7. Birkhoff, G., *Lattice Theory*, 3rd edition, volume 25 of AMS Colloquium Publications, Providence, 1967.

8. Borsuk, K., Sur la décomposition des polyèdres en produits cartésiens, Fund. Math. 31 (1938), 137-148.

9. Bulman-Fleming, S., and W. Taylor, Union-indecomposable varieties, Colloq. Math. 35 (1976), 189-199.

10. *Burnside Groups*, Proceedings 1977, Lecture Notes 806 Springer, Edited by J. L. Mennike, 1980.

11. Burris, S., Models in equational theories of unary algebras, Algebra Universalis 1 (1971), 386-392.

12. Burris, S., and H.P. Sankappanavar, *A Course in Universal Algebra*, Springer-Verlag, Berlin, 1980.

13. Chajda, I., Coherence, regularity and permutability of congruences, Algebra Universalis, to appear.

14. Clifford, A., and G. Preston, *The Algebraic Theory of Semigroups I*, AMS, 1964.

15. Csákány, B., Varieties whose algebras have no idempotent elements, Colloq. Math. 35 (1976), 201-203.

16. ———— Selective algebras and compatible varieties (to appear).

17. Davey, B., and H. Werner, Dualities and equivalences for varieties of algebras, Colloq. Math Soc. János Bolyai. 33 (1983), 181-275.

18. Day, A., A characterization of modularity for congruence lattices of algebras, Canad. Math. Bull. 12 (1969), 167-173.

19. ———— A simple solution to the word problem for lattices, Canad. Math. Bull. 13 (1970), 253-254.

20. Dean, R.A., and T. Evans, A remark on varieties of lattices and semigroups, AMS Proceedings 21 (1969), 394-396.

21. Felscher, W., Equational maps, 121-161 in: H. A. Schmidt, K. Schütte and H.-J. Thiele, eds., *Contributions to Mathematical Logic*, North-Holland, Amsterdam, 1968.

22. Freese, R., The variety of modular lattices is not generated by its finite members, AMS Transactions 255 (1979), 277-300.

23. Freese, R., and J.B. Nation, Congruence lattices of semilattices, Pacific J. Math. 49 (1973), 51-58.

24. Freyd, P.J., Algebra-valued functors in general and tensor products in particular, Colloq. Math. 14 (1966), 89-106.

25. Froemke, J. and R.W. Quackenbush, The spectrum of an equational class of groupoids, Pacific J. Math. 58 (1975), 381-386.

26. Galvin, F., Horn sentences, Annals Math. Logic 1 (1970), 389-422.

27. García, O.C., Grupos de exponente 4: Subvariedades de A_2A_2, Anales del Instituto de Matemáticas (UNAM, México) 17 (1977), 41-74.

28. García, O.C., F. Larrión and W. Taylor, On the lattice of interpretability types of varieties, Anales del Instituto de Matemáticas (UNAM, México) 22 (1982), 189-195.

29. García, O.C., and J.A. de la Peña, Lattices with a finite Whitman cover, Algebra Universalis 16 (1983), 186-194.

30. García, O.C. and M. Solay, A classification in varieties of algebras, Publicaciones Preliminares del Instituto de Matemáticas (UNAM, México) 21 (1980), 1-12.

31. Glimm, J., Two cartesian products which are Euclidean spaces, Bull. Soc. Math. France 88 (1960), 131-135.

32. Grätzer, G., On the spectra of classes of algebras, AMS Proceedings 18 (1967), 729-735.

33. ———— *Universal Algebra*, Van Nostrand, Princeton, 1968. Second edition, Springer, 1979.

34. ———— Composition of functions, 1-106 in the Proceedings of the Conference on Universal Algebra at Queen's University, Kingston, Ontario, 1969. Volume 25 of the Queen's Papers in Pure and Applied Mathematics.

35. ———— *Lattice Theory*, First Concepts and Distributive Lattices, H.M. Freeman, San Francisco, 1971.

36. ———— *General Lattice Theory*, Birkhäuser, Basel, 1978.

37. Gumm, H.-P., Mal'cev conditions in sums of varieties and a new Mal'çev condition, Algebra Universalis 5 (1975), 56-64.

38. ———— Congruence modularity is permutability composed with distributivity, Arch. Math. (Basel) 36 (1981), 569-576.

39. Hall, M. Jr., *Combinatorial Theory*, Blaisdell, Waltham, 1967.

40. Hall, M. Jr., and J.K. Senior, *The Groups of Order* 2^n $(n \leq 6)$, The MacMillan Co., 1964.

41. Isbell, J.R., Compatibility and extensions of algebraic theories, Algebra Universalis 6 (1976), 37-51.

42. Jech, T.J., *Lectures in Set Theory*, volume 217 of Lecture Notes in Mathematics.

43. Jónsson, B., Algebras whose congruence lattices are distributive, Math. Scand. 21 (1967), 110-121.

44. ———— Congruence varieties, Appendix 3 to Grätzer [33], pp. 348-378.

45. ———— Congruence varieties, Algebra Universalis 10 (1980) 355-394.

46. Keisler, H.J., Reduced products and Horn classes, AMS Transactions 117 (1965), 307.

47. Kelley, G.M., and A. Pultr, On algebraic recognition of direct-product decompositions, J. Pure and Applied Algebra 12 (1978), 207-224.

48. Kollár, J., Congruences and one-element subalgebras, Algebra Universalis 9 (1979), 266-267.

49. Kreisel, G., Some uses of metamathematics, British J. Philos. Sci. 7 (1956), 161.

50. Kwun, K.W., and F. Raymond, Factors of cubes, Amer. J. Math. 84 (1962), 433-440.

51. Lausch, H., and W. Nöbauer, *Algebra of Polynomials*, North-Holland, Amsterdam, 1973.

52. Levi, F., and B.L. van der Waerden, Über eine besondere Klasse von Gruppen, Abh. aus dem Math. Sem. der Hamburg Univ. 9 (1932), 154-158.

53. Mal'tsev, A.I., The structural characterization of certain classes of algebras (in Russian), Dokl. Akad. Nauk SSSR 120 (1958), 29-32. Translated as Chapter 9 of B.F. Wells (ed.), *The Metamathematics of Algebraic Systems* (Collected papers of A.I. Mal'tsev, 1936-67), North-Holland, Amsterdam, 1971.

54. McKenzie, R., Negative solution of the decision problem for sentences true in every subalgebra of $\langle N,+\rangle$, J. Symbolic Logic 36 (1971), 607-609.

55. ———— Equational bases for lattice theories, Math. Scand. 27 (1970), 24-38.

56. ———— On spectra, and the negative solution of the decision problem for identities having a non-trivial finite model, J. Symbolic Logic 40 (1975), 186-196.

57. McNulty, G., Undecidable properties of finite sets of equations, J. Symbolic Logic 41 (1976), 9-15.

58. Meier-Wunderli, H., Metabelsche Gruppen, Comment. Math. Helvet. 25 (1951), 1-10.

59 van Mill, J., A rigid space X for which $X \times X$ is homogeneous: an application of infinite-dimensional topology, AMS Proceedings 83 (1981), 597-600.

60. Mycielski, J., A lattice of interpretability types of theories, J. Symbolic Logic 42 (1977), 297-305.

61. Nagami, K., *Dimension Theory*, Academic Press, New York, 1970.

62. Neumann, H., *Varieties of Groups*, Ergebnisse der Mathematik Vol. 37, Springer Verlag, 1967.

63. Neumann, W.D., On Mal'cev conditions, J. Austral. Math. Soc. 17 (1974), 376-384.

64. ———— Mal'cev conditions, spectra and Kronecker product, J. Austral. Math. Soc. (A) 25 (1978), 103-117.

65. Newman, M.H.A., and J.H.C. Whitehead, On the group of a certain linkage, Quart. J. Math. 8 (1937), 14-21.

66. de la Peña, J.A., On interpretability in varieties of groups, Algebra Universalis, to appear.

67. Perkins, P., Unsolvable problems for equational theories, Notre Dame J. Formal Logic 8 (1967), 175-185.

68. Quackenbush, R.W., Primality: the influence of Boolean algebras in universal algebra, Appendix 5 to Grätzer [33], pp. 401-416.

69. ———————— Varieties with n-principal compact congruences, Algebra Universalis 14 (1982), 292-296.

70. Sachs, D., Identities in finite partition lattices, AMS Proceedings 12 (1961), 944-945.

71. Taylor, W., Fixed points of endomorphisms, Algebra Universalis 2 (1972), 74-76.

72. ———— Characterizing Mal'tsev conditions, Algebra Universalis 3 (1973), 351-397.

73. ———— Uniformity of congruences, Algebra Universalis 4 (1974), 342-360.

74. ———— The fine spectrum of a variety, Algebra Universalis 5 (1975), 263-303.

75. ———— Pure compactifications in quasi-primal varieties, Canad. J. Math. 28 (1976), 50-62.

76. ———— Varieties obeying homotopy laws, Canad. J. Math. 29 (1977), 498-527.

77. ———— *Equational Logic*, Houston J. Math., survey issue, 1979, iii + 83 pp.

78. ———— Primal topological algebras, Algebra Universalis 9 (1979), 211-220.

79. ———— Mal'tsev conditions and spectra, J. Austral. Math. Soc. (A) 29 (1980), 143-152.

80. Thurston, H.A., Derived operations and congruences, Proc. London Math. Soc. (3) 8 (1958), 127-134.

O. C. Garcia
Instituto de Matemáticas
Universidad Nacional Autónoma de México
México, D.F. 04510
MEXICO

W. Taylor
Mathematics Department
University of Colorado
Boulder, CO 80309-0426
USA

General instructions to authors for
PREPARING REPRODUCTION COPY FOR MEMOIRS

> **For more detailed instructions send for AMS booklet, "A Guide for Authors of Memoirs." Write to Editorial Offices, American Mathematical Society, P. O. Box 6248, Providence, R. I. 02940.**

MEMOIRS are printed by photo-offset from camera copy fully prepared by the author. This means that, except for a reduction in size of 20 to 30%, the finished book will look exactly like the copy submitted. Thus the author will want to use a good quality typewriter with a new, medium-inked black ribbon, and submit clean copy on the appropriate model paper.

Model Paper, provided at no cost by the AMS, is paper marked with blue lines that confine the copy to the appropriate size. Author should specify, when ordering, whether typewriter to be used has PICA-size (10 characters to the inch) or ELITE-size type (12 characters to the inch).

Line Spacing – For best appearance, and economy, a typewriter equipped with a half-space ratchet – 12 notches to the inch – should be used. (This may be purchased and attached at small cost.) Three notches make the desired spacing, which is equivalent to 1-1/2 ordinary single spaces. Where copy has a great many subscripts and superscripts, however, double spacing should be used.

Special Characters may be filled in carefully freehand, using dense black ink, or INSTANT ("rub-on") LETTERING may be used. AMS has a sheet of several hundred most-used symbols and letters which may be purchased for $5.

Diagrams may be drawn in black ink either directly on the model sheet, or on a separate sheet and pasted with rubber cement into spaces left for them in the text. Ballpoint pen is *not* acceptable.

Page Headings (Running Heads) should be centered, in CAPITAL LETTERS (preferably), at the top of the page – just above the blue line and touching it.

LEFT-hand, EVEN-numbered pages should be headed with the AUTHOR'S NAME;

RIGHT-hand, ODD-numbered pages should be headed with the TITLE of the paper (in shortened form if necessary).

Exceptions: PAGE 1 and any other page that carries a display title require NO RUNNING HEADS.

Page Numbers should be at the top of the page, on the same line with the running heads.

LEFT-hand, EVEN numbers – flush with left margin;

RIGHT-hand, ODD numbers – flush with right margin.

Exceptions: PAGE 1 and any other page that carries a display title should have page number, centered below the text, on blue line provided.

FRONT MATTER PAGES should be numbered with Roman numerals (lower case), positioned below text in same manner as described above.

MEMOIRS FORMAT

> **It is suggested that the material be arranged in pages as indicated below.**
> **Note: Starred items (*) are requirements of publication.**

Front Matter (first pages in book, preceding main body of text).

Page i – *Title, *Author's name.

Page iii – Table of contents.

Page iv – *Abstract (at least 1 sentence and at most 300 words).

*1980 Mathematics Subject Classifications represent the primary and secondary subjects of the paper. For the classification scheme, see Annual Subject Indexes of MATHEMATICAL REVIEWS beginning in December 1978.

Key words and phrases, if desired. (A list which covers the content of the paper adequately enough to be useful for an information retrieval system.)

Page v, etc. – Preface, introduction, or any other matter not belonging in body of text.

Page 1 – Chapter Title (dropped 1 inch from top line, and centered).

Beginning of Text.

Footnotes: *Received by the editor date.

Support information – grants, credits, etc.

Last Page (at bottom) – Author's affiliation.

ABCDEFGHIJ–AMS–8987654

DATE DUE

DEMCO 38-297

OCT 1 1984